Padé Approximants

Part II: Extensions and Applications

GIAN-CARLO ROTA, *Editor*
ENCYCLOPEDIA OF MATHEMATICS AND ITS APPLICATIONS

GIAN-CARLO ROTA, *Editor*
ENCYCLOPEDIA OF MATHEMATICS AND ITS APPLICATIONS

Other volumes in preparation

ENCYCLOPEDIA
OF MATHEMATICS
and Its Applications

GIAN-CARLO ROTA, Editor
Department of Mathematics
Massachusetts Institute of Technology
Cambridge, Massachusetts

Editorial Board

GIAN-CARLO ROTA, *Editor*

ENCYCLOPEDIA OF MATHEMATICS AND ITS APPLICATIONS

Volume 14

Section: Mathematics of Physics
Peter A. Carruthers, *Section Editor*

Padé Approximants
Part II: Extensions and Applications

George A. Baker, Jr.
Los Alamos National Laboratory
Los Alamos, New Mexico

Peter Graves-Morris
University of Kent
Canterbury, Kent, England

Foreword by
Peter A. Carruthers
Los Alamos National Laboratory

1981

Addison-Wesley Publishing Comapny
Advanced Book Program
Reading, Massachusetts

London · Amsterdam · Don Mills, Ontario · Sydney · Tokyo

Library of Congress Cataloging in Publication Data

Baker, George A. (George Allen), Jr., 1932–
Padé approximants.

(Encyclopedia of mathematics and its applica-
tions; 13–14. Section, Mathematics of physics)
Bibliography: p.
Includes index.
Contents: pt. 1. Basic theory—pt. 2. Extensions
and applications.
1. Padé approximant. I. Graves-Morris, P. R.
II. Title. III. Series: Encyclopedia of mathe-
matics and its applications; v. 13–14.
IV. Series: Encyclopedia of mathematics and its applications.
Section, Mathematics of physics.
QC20.7.P3B35   530.1′514   81-3546
ISBN 0-201-13512-4 (v. 1)   AACR2
ISBN 0-201-13513-2 (v. 2)

American Mathematical Society (MOS) Subject Classification Scheme (1980): 41A21

Copyright ©1981 by Addison-Wesley Publishing Company, Inc.
Published simultaneously in Canada.

Manufactured in the United States of America

ABCDEFGHIJ–HA–8987654321

To Our Wives
Elizabeth Baker *and* **Lucia Graves-Morris**
and to Our Families

Contents of Part II

Contents of Part I

Editor's Statement

A large body of mathematics consists of facts that can be presented and described much like any other natural phenomenon. These facts, at times explicitly brought out as theorems, at other times concealed within a proof, make up most of the applications of mathematics, and are the most likely to survive changes of style and of interest.

This ENCYCLOPEDIA will attempt to present the factual body of all mathematics. Clarity of exposition, accessibility to the non-specialist, and a thorough bibliography are required of each author. Volumes will appear in no particular order, but will be organized into sections, each one comprising a recognizable branch of present-day mathematics. Numbers of volumes and sections will be reconsidered as times and needs change.

It is hoped that this enterprise will make mathematics more widely used where it is needed, and more accessible in fields in which it can be applied but where it has not yet penetrated because of insufficient information.

GIAN-CARLO ROTA

Foreword

New insights into mathematical problems and physical applications continue to arise from the study of power series representation of functions. The present work is devoted to an intensive description of the Padé approximant technique*. The value of this scheme of approximation in a wide variety of physical problems has been increasingly recognized in recent years. This two-part presentation is a fine example of the interplay between physics and mathematics, each stimulating the other to new concepts and techniques.

One could not imagine better qualified authors for a contemporary major set of volumes on Padé approximants. Baker and Graves-Morris are widely known for their original contributions both to the mathematics and serious physical applications. The result is a lucid and explicit treatment of the subject which does not compromise mathematical accuracy yet is easily accessible to the modern theorist.

We may mention that this work is an example of a healthy trend developing in recent years in which modern mathematical developments are increasingly providing the language in which the most advanced physical theories are expressed. In the present case the renaissance in statistical mechanics and field theory studies in recent years has required such developments as Wilson's renormalization group method and Padé approximants. We may also mention the serious studies of continuous groups and their representations inspired by efforts to unite the weak, electromagnetic, strong, and gravitational forces. These same theories seem to be best formulated as non-Abelian gauge field theories, whose content and consequences involve the concepts of differential geometry and topology.

Stated briefly, the Padé approximant represents a function by the ratio of two polynomials. The coefficients of the powers occurring in the polynomials are, however, determined by the coefficients in the Taylor series expansion of the function. Thus, given a power series expansion

$$f(z) = c_0 + c_1 z + c_2 z^2 \ldots$$

we can by the methods described in the text, make an optimal choice of the

---

*Among other related techniques we mention continued fractions treated in volume 11 of this Encyclopedia (Jones and Thron).

coefficients a_i, b_i in the Padé approximant

$$\frac{a_0 + a_1 z + \cdots + a_L z^L}{b_0 + b_1 z + \cdots + b_M z^M}$$

Exploitation of this simple idea and its extensions has led to many insights and by now has become practically a major industry. I shall not spoil the story by revealing more of the plot.

Inspection of the table of contents reveals an intensive development of the mathematical texture inherent in the subject. Many excellent examples illustrate the concepts. Some recent results appear here for the first time in monograph form. Among these are included the developments of reliable algorithms (I.2.1, 2.4, 4.5, and II.1.1), Saff-Varga theorems on Padé approximants to the exponential (I.5.7), the theory of convergence in capacity of Pommerenke (I.6.6), Canterbury (two variable) approximants (II.1.4) and results from $\lambda\phi^4$ Euclidean field theory derived using Padé approximants. In addition the approach to Laguerre's method in (II.3.7) is new. The treatment of applications is well done and has sufficient depth to be useful to the research scientist. Part II, Chapter 2, describes the connection with integral equations and quantum mechanics. The connection with numerical analysis is made in Part II, Chapter 3. The authors close with a frontier topic, the application of Padé approximants to problems in quantum field theory. Finally, an extensive bibliography documents the subject and provides references to the treatment of further related problems.

This two-volume presentation is a fine example of a creative review because it weaves together the vital ideas of the subject of Padé approximants and sets the stage for vigorous new developments in theory and applications. It should fill this role for some time to come.

PETER A. CARRUTHERS
General Editor, Section on Mathematics of Physics

Preface

These two volumes are intended to serve as a basic text on one approach to the problem of assigning a value to a power series. We have attempted to present the basic results and methods in as transparent a form as possible, in line with the general objectives of the Encyclopedia. The general topic of Padé Approximants, which is, among other things, a highly practical method of definition and of construction of the value of a power series, seems to have begun independently at least twice. Padé's claim for credit is based on his thesis (1892), in which he developed the approximants and organized them in a table. He paid particular attention to the exponential function. He was presumably unaware of the prior work of Jacobi (1846), who gave the determinantal representation in his paper on the simplification of Cauchy's solution to the problem of rational interpolation. Also, Padé's work was preceded by that of Frobenius (1881), who derived identities between the neighboring rational fractions of Jacobi. It is interesting to note that Anderson seems to have stumbled upon some Padé Approximants for the logarithmic function in 1740. A photograph of H. Padé is to be found in *The Padé Approximant Method and Its Application to Mechanics*, edited by H. Cabannes. A copy of his autographed thesis is to be found in the Cornell University Library.

This work has been distilled from an extensive literature, and *The Essentials of Padé Approximants*, written by one of us, has been an essential reference. We use the abbreviation EPA for this book, and refer to it often for a different or fuller treatment of some of the more advanced topics. While each book is entirely self-contained, our notation is normally compatible with EPA, and to a large extent the books complement each other. An important exception is that the Padé table in EPA is reflected through its main diagonal in our present notation. The proceedings of the Canterbury Summer School and International Conference, edited by the other of us, contain diverse contributions which initiated in print the multidisciplinary view of the subject—a view we hope we have transmitted herein. The many publications which have contributed substantially to our text are listed in the bibliography. We are grateful to our numerous colleagues at Brookhaven, Canterbury, Cornell, Los Alamos, and Saclay in freely discussing so many topics which have made possible the breadth of our treatment. Especially, we thank Roy Chisholm, John Gammel, and Daniel Bessis for many conversations, and the C.E.A. at Saclay, where part of this book was written, for hospitality.

Our hardest task in writing this book was to choose a presentation which is both correct and readily comprehensible. A fully precise system based on rigorous analysis and set-theoretic language would have ensured total obscurity of the more practical techniques. Conversely, omission of all the conditions under which the theorems hold good would be absurdly misleading. We have chosen a level of presentation suitable for the topic in hand. For example, the connectivity of sets is mentioned where it is important, and otherwise it is omitted. The meaning of the order notation is clear in context. Both applications in physics and techniques recently developed are treated in a practical fashion.

Equations are referenced by a default option. Equation (I.6.5.3) is Equation (5.3) of Part I, Chapter 6; the Part and Chapter are dropped by default if they are the same as the source of the reference.

Finally, a spirit of evangelism may be detected in the text. When a review of rational approximation in 1963 can claim that Padé approximants cannot approximate on the entire range $(0, \infty)$ and be believed, a revision of view is overdue.

GEORGE A. BAKER, JR.
PETER GRAVES-MORRIS

Padé Approximants

Part II: Extensions and Applications

CHAPTER 1

Extensions of Padé Approximants

1.1. Multipoint Padé Approximants

A rational function which fits given function values at various points, not necessarily distinct, is called a multipoint Padé approximant. The associated problem of interpolation by rational functions is called the Cauchy–Jacobi problem. Multipoint Padé approximants are also called rational inter-polants, N-point Padé approximants, or Newton Padé approximants, de-pending on the context. Interpolation at confluent points is sometimes called osculatory interpolation. For example, an appealing idea is that the $[N/N]$ rational form should satisfy N derivative conditions at $z=0$ and $N-1$ derivative conditions at $z=\infty$. Specifically, consider

$$f(z)=\left(\frac{1+\frac{1}{2}z}{1+2z}\right)^{1/2}$$

$$=1-\tfrac{3}{4}z+O(z^2) \qquad \text{as} \quad z\to 0, \tag{1.1}$$

$$=\tfrac{1}{2}+O(z^{-1}) \qquad \text{as} \quad z\to \infty. \tag{1.2}$$

Linear algebra shows that the rational approximant so defined is

$$R^{[1/1]}(z)=\frac{1+\frac{3}{4}z}{1+\frac{3}{2}z},$$

which satisfies the accuracy-through-order conditions indicated by (1.1) and (1.2). At $z=1$, the accuracy is 1%, which is the same absolute accuracy as is achieved by the [1/1] Padé approximant (cf. Part I, Section 1.1).

ENCYCLOPEDIA OF MATHEMATICS and Its Applications, Gian-Carlo Rota (ed.).
Vol. 14: George A. Baker, Jr., and Peter R. Graves-Morris, Padé Approximants, Part II:
Extensions and Applications   ISBN 0-201-13513-2

This analysis of a particular case indicates the approach to the general problem of rational interpolation at $z_0, z_1, z_2, \ldots$. The previous example is about 3-point interpolation with $z_0 = z_1 = 0$ and $z_2 = \infty$. The analysis of the general case is easier using finite points, and we are led to consider rational interpolation allowing confluence, and associated approximation problems. Space does not permit a full account of all these problems; we begin with an introduction to polynomial and rational interpolation, with particular emphasis on methods which allow confluence. First, we need the basic framework of Newtonian polynomial interpolation.

DIVIDED DIFFERENCES For a function $f(z)$ satisfying such continuity properties as are necessary, we define

$$f[z_0] = f(z_0), \tag{1.3a}$$

$$f[z_0, z_1] = \frac{f(z_0) - f(z_1)}{z_0 - z_1}, \tag{1.3b}$$

and other divided differences are defined recursively by

$$f[z_0, z_1, \ldots, z_{r+1}] = \frac{f[z_0, z_1, \ldots, z_{r-1}, z_r] - f[z_0, z_1, \ldots, z_{r-1}, z_{r+1}]}{z_r - z_{r+1}},$$

$$r = 1, 2, \ldots . \tag{1.4}$$

HERMITE'S FORMULA. If $f(z)$ is analytic inside and continuous on a contour Γ enclosing $z_0, z_1, \ldots, z_k$, then

$$f[z_0, z_1, \ldots, z_r] = \frac{1}{2\pi i} \int_\Gamma \frac{f(\zeta)}{\displaystyle\prod_{k=0}^{r} (\zeta - z_k)} \, d\zeta \tag{1.5}$$

Proof. The proof is by induction using (1.3) and (1.4).

For confluent points $z_0 = z_1 = \cdots = z_r$, it is natural to define

$$f[z_0, z_0, \ldots, z_0] = \frac{f^{(r)}(z_0)}{r!}. \tag{1.6}$$

Hermite's formula easily extends to cases of partial confluence.

COROLLARY. $f[z_0, z_1, \ldots, z_r]$ is a totally symmetric function of all its arguments $z_0, z_1, \ldots, z_r$.

NEWTON'S FORMULAS.

$$f(z) = \sum_{i=0}^{n} f[z_0, z_1, \ldots, z_i] \prod_{k=0}^{i-1} (z - z_k)$$

$$+ f[z_0, z_1, \ldots, z_n, z] \prod_{k=0}^{n} (z - z_k). \qquad (1.7)$$

For $n > 0$, (1.7) is an identity expressing $f(z)$ as a Newton polynomial and a remainder term. One may "deduce" the formal identity

$$f(z) = f[z_0] + (z - z_0) f[z_0, z_1] + (z - z_0)(z - z_1) f[z_0, z_1, z_2] + \cdots. \qquad (1.8)$$

Whenever the remainder in (1.7) tends to zero, (1.8) becomes an identity. The proof of (1.7) by induction is straightforward. It is the interpretation of (1.7) and (1.8) that is most significant. If $z_0 = z_1 = \cdots = z_i = \cdots$, (1.7) and (1.8) become

$$f(z) = f(z_0) + (z - z_0) f'(z_0) + \frac{(z - z_0)^2}{2!} f''(z_0) + \cdots \qquad (1.9)$$

$$= \sum_{i=0}^{n} \frac{(z - z_0)^i}{i!} f^{(i)}(z_0) + \frac{(z - z_0)^{n+1}}{2\pi i} \int_{\Gamma} \frac{f(\zeta)\, d\zeta}{(\zeta - z_0)^n (\zeta - z)}. \qquad (1.10)$$

Equation (1.10) holds provided Γ is a contour enclosing z, z_0 and $f(z)$ is analytic within Γ and continuous on Γ. In fact, (1.10) gives the Taylor series for $f(z)$ and its remainder. For conciseness, we make a further definition:

DEFINITION

$$f_{i,j} = f[z_i, z_{i+1}, \ldots, z_j], \qquad \text{for} \quad j \geq i. \qquad (1.11)$$

Then Newton's formula (1.8) becomes the formal identity

$$f(z) = f(z_0) + f_{0,1}(z - z_0) + f_{0,2}(z - z_0)(z - z_1) + \ldots.$$

We now proceed to consider interpolation of a given function $f(z)$ using rational fractions which are sometimes called interpolants. The basic problem is to find a rational fraction

$$r^{[L/M]}(z) = u^{[L/M]}(z) / v^{[L/M]}(z) \qquad (1.12)$$

where $u^{[L/M]}(z)$ has maximum order L, $v^{[L/M]}(z)$ has maximum order M

and

$$r^{[L/M]}(z_i)=f(z_i), \qquad i=0,1,2,\ldots,L+M. \tag{1.13}$$

If a solution to this basic problem exists, it is obtained by defining

$$u^{[L/M]}(z)= \sum_{j=0}^{L} u_j z^j, \qquad v^{[L/M]}(z)= \sum_{k=0}^{M} v_k z^k \tag{1.14}$$

for specific values of L, M. Let us assume that $v_0=1$ is a permissible normalization for the moment. Substitution of (1.12) and (1.14) into (1.13) yields $L+M+1$ linear equations for $L+M+1$ unknown coefficients $u_0, u_1,\ldots, u_L, v_1, v_2,\ldots, v_M$. Normally there is a unique solution leading to a rational interpolant which is uniquely defined up to a constant common factor in the numerator and denominator of (1.12). Otherwise, the equations are said to be degenerate. If the equations are degenerate but consistent, and $v^{[L/M]}(z)\not\equiv 0$, then $u^{[L/M]}(z)$ and $v^{[L/M]}(z)$ have a common factor. Using (1.19) with $f(z)=r^{[L/M]}(z)$, it follows that the factors $(z-z_i)$, $i=0,1,\ldots,$ $L+M$ are the only possible elementary common factors of $u^{[L/M]}(z)$ and $v^{[L/M]}(z)$. For each such factor $(z-z_k)$, (1.13) must be tested with $i=k$ for the proposed solution. If the linear equations are inconsistent, no rational fraction of type $[L/M]$ fits the data. As an example, we next show that no rational function of type $[1/1]$ fits the data

$$f(-1)=1, \qquad f(0)=1, \qquad f(1)=3 \tag{1.15}$$

at the indicated points. The equations (1.12), (1.13), and (1.14) become

$$u_0-u_1=v_0-v_1, \tag{1.16a}$$

$$u_0=v_0, \tag{1.16b}$$

$$u_0+u_1=3(v_0+v_1). \tag{1.16c}$$

Equation (1.16a, b) imply that $u_0=v_0$, $u_1=v_1$, and so (1.16c) implies that $u_0=u_1=v_0=v_1=0$. The equations (1.16) are degenerate. In fact, only the new value $f(1)=1$ would render (1.16) consistent and allow rational interpolation to be effected by a (degenerate) interpolant of type $[1/1]$. A full analysis of possible degeneracies is given by Maehly and Witzgall [1960].

Since Padé approximation is rational approximation with complete confluence of the interpolation points, it is interesting to note the similarity between the previous analysis and that of the existence or nonexistence of Padé approximants in Part I, Section 1.4. Having briefly considered some of the hazards of using rational interpolation, the next theorem gives the standard solution in the nondegenerate case.

THEOREM 1.1.1. *The N-point Padé approximant of type* $[L/M]$ *defined by interpolation at the points* $z_0, z_1, \ldots, z_{L+M}$, *allowing confluence, is normally given by*

$$r^{[L/M]}(z) = u^{[L/M]}(z)/v^{[L/M]}(z),$$

where $u^{[L/M]}(z)$ *and* $v^{[L/M]}(z)$ *are defined by*

$$u^{[L/M]}(z) =$$

$$\times \begin{vmatrix} f_{M,L+1} & f_{M,L+2} & \cdots & f_{M,L+M} & \sum_{j=M}^{L} f_{M,j} \prod_{k=0}^{j-1} (z-z_k) \\ f_{M-1,L+1} & f_{M-1,L+2} & \cdots & f_{M-1,L+M} & \sum_{j=M-1}^{L} f_{M-1,j} \prod_{k=0}^{j-1} (z-z_k) \\ \vdots & \vdots & & \vdots & \vdots \\ f_{0,L+1} & f_{0,L+2} & \cdots & f_{0,L+M} & \sum_{j=0}^{L} f_{0,j} \prod_{k=0}^{j-1} (z-z_k) \end{vmatrix},$$

$$(1.17)$$

$$v^{[L/M]}(z) =$$

$$\begin{vmatrix} f_{M,L+1} & f_{M,L+2} & \cdots & f_{M,L+M} & \prod_{k=0}^{M-1} (z-z_k) \\ f_{M-1,L+1} & f_{M-1,L+2} & \cdots & f_{M-1,L+M} & \prod_{k=0}^{M-2} (z-z_k) \\ \vdots & \vdots & & \vdots & \vdots \\ f_{0,L+1} & f_{0,L+2} & \cdots & f_{0,L+M} & 1 \end{vmatrix}, \qquad (1.18)$$

and the definition (1.11) *has been used.*

The remainder is given by

$$v^{[L/M]}(z)f(z) - u^{[L/M]}(z) = \prod_{k=0}^{L+M} (z-z_k)$$

$$\times \begin{vmatrix} f_{M,L+1} & f_{M,L+2} & \cdots & f_{M,L+M} & f[z_M,\ldots,z_{L+M},z] \\ f_{M-1,L+1} & f_{M-1,L+2} & \cdots & f_{M-1,L+M} & f[z_{M-1},\ldots,z_{L+M},z] \\ \vdots & \vdots & & \vdots & \vdots \\ f_{0,L+1} & f_{0,L+2} & \cdots & f_{0,L+M} & f[z_0,\ldots,z_{L+M},z] \end{vmatrix}.$$

$$(1.19)$$

If "impossible" entries in (1.17)–(1.19) occur, the following interpretation is intended: If $j < i$, then

$$f_{i,j} = 0, \qquad \sum_{k=i}^{j} (term)_k = 0, \quad and \quad \prod_{k=i}^{j} (factor)_k = 1.$$

A sufficient condition for the result that $u^{[L/M]}(z_i)/v^{[L/M]}(z_i) = f(z_i)$ is that $v^{[L/M]}(z_i) \neq 0$, $i = 0, 1, \ldots, L+M$.

Proof. The formulas (1.17) and (1.18) are polynomials of the appropriate orders. Using Newton's formula (1.7), it follows that

$$v^{[L/M]}(z)f(z) - u^{[L/M]}(z) = \prod_{k=0}^{L} (z - z_k) \times$$

$$\begin{vmatrix} f_{M,L+1} & f_{M,L+2} & \cdots & f_{M,L+M} & f[z_M, \ldots, z_L, z] \\ f_{M-1,L+1} & f_{M-1,L+2} & \cdots & f_{M-1,L+M} & f[z_{M-1}, \ldots, z_L, z] \\ \vdots & \vdots & & \vdots & \vdots \\ f_{0,L+1} & f_{0,L+2} & \cdots & f_{0,L+M} & f[z_0, \ldots, z_L, z] \end{vmatrix}.$$

$$(1.20)$$

Recalling the definitions (1.4) and (1.11), repeated subtraction of the jth column of (1.20) from the last for $j = 1, 2, \ldots, M$ yields (1.19). This is manifestly zero at $z_0, \ldots, z_{L+M}$. Provided $v^{[L/M]}(z_i) \neq 0$, $i = 0, 1, \ldots, L+M$, the result follows.

COROLLARY. If $z_0 = z_1 = \cdots = z_{L+M}$, (1.6) shows that $r^{[L/M]}(z)$ defined by (1.12), (1.17), and (1.18) reduces to a Padé approximant as in (I.1.1.8, 9, 11).

NOTE. Eqns. (1.17), (1.18), (1.19) can be generalized to solve the bigradient problem (Part I, Section 1.6) with interpolation at distinct points [Warner, 1974; Householder and Stewart, 1969].

Construction of the N-point Padé approximants using (1.17), (1.18) is cumbersome and quite unsuitable for numerical work. We now turn to consider some of the requirements of a satisfactory numerical algorithm.

Algorithms for rational interpolation may be classified into those designed to solve the coefficient problem and those designed to solve the value problem. The coefficient problem is the problem of evaluating the coefficients $\{u_0, u_1, \ldots, u_L; v_0, v_1, \ldots, v_M\}$ in (1.14) and thereby defining the rational form $r^{[L/M]}(z)$ in (1.12). The value problem entails evaluation of $r^{[L/M]}(z)$ at some prespecified value of z without necessarily finding the coefficients in (1.14) explicitly. For example, the ϵ-algorithm is a method of solving a value problem for Padé approximants without construction of the coefficients. The Q.D. algorithm for constructing continued fractions (see Part I, Section 4.4) is the solution of a coefficient problem.

If a table of values of an interpolant is required, it is usually more economical to solve the coefficient problem first and then to evaluate the interpolant at the various values of z. If only a single value of an interpolant is required, it is sometimes more efficient not to evaluate the coefficients explicitly. In fact, evaluation of continued fractions and polynomials is a comparatively rapid process. Consequently, the coefficient problem is especially important. Notice that computational economy is gained by using continued-fraction representations of interpolants of the type $r^{[M/M]}(z)$ rather than using polynomial ratios (1.12).

An important and desirable property of rational interpolation methods is the property of allowing confluence of some or all of the interpolation points. Methods based directly on divided differences (1.3), (1.4) are intrinsically incapable of generalization to osculatory interpolation.

Reliability of rational interpolation methods is another quality we seek. As stated earlier, no [1/1]-type interpolant fits the data of (1.15), and consequently any reliable interpolation method should recognize this problem as insoluble. Rational interpolation algorithms must discriminate between soluble and insoluble problems, making due allowances for rounding and representation errors. This consideration leads us to the property of stability, which is closely related to reliability. If an algorithm is stable, small changes in the data lead to small changes in the results. A good rational interpolation algorithm should be able to detect data which give unstable results.

We observe that sequential methods of constructing a particular rational interpolant depend on the existence of all the intermediate interpolants of the sequence. A reliable algorithm will work when the desired interpolant exists, even if the intermediate interpolants are degenerate or do not exist.

The qualities characterizing a good algorithm are notorious for not being entirely compatible, and the selection of a "best algorithm" involves a compromise. For each of the various rational interpolation problems, we seek algorithms which (i) are efficient, (ii) allow confluence, and (iii) are stable and reliable. We next turn to consider some of the best of the algorithms available.

KRONECKER'S ALGORITHM [Kronecker, 1881]. To start the algorithm, the Newton interpolating polynomial (1.8) of order $L+M=m$ is used, with Newton coefficients derived using (1.4). Thus the initializing values are the coefficients in

$$p^{[m/0]}(z) = \sum_{i=0}^{m} f[z_0, z_1, \ldots, z_i] \prod_{k=0}^{i-1} (z - z_k), \qquad (1.21a)$$

$$q^{[m/0]}(z) = 1, \qquad (1.21b)$$

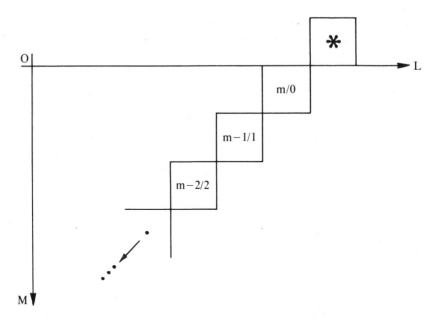

Figure 1.   Elements of the N-point Padé table used in Kronecker's algorithm.

and the artificial initializing values are given by

$$p^{[m+1/-1]}(z) = \prod_{k=0}^{m} (z - z_k), \qquad (1.22a)$$

$$q^{[m+1/-1]}(z) = 0. \qquad (1.22b)$$

The recurrence relations (which we will prove) are analogues of (I.3.6.12) and are

$$p^{[m-j/j]}(z) = (\alpha_j z + \beta_j) p^{[m-j+1/j-1]}(z) - p^{[m-j+2/j-2]}(z), \qquad (1.23a)$$

$$q^{[m-j/j]}(z) = (\alpha_j z + \beta_j) q^{[m-j+1/j-1]}(z) - q^{[m-j+2/j-2]}(z). \qquad (1.23b)$$

At stage (j), Equation (1.23a) is used to determine both α_j and β_j, so that the apparent degree of the right-hand side of (1.23a) is reduced by two. Then (1.23a) gives the next numerator and (1.23b) the next denominator.

Verification. We assume that the algorithm is nonsingular, which is tantamount to assuming that the numerators have $p^{[m-k/k]}$ have the full indicated order ($m-k$) for all values of k needed. The coefficients α_j and β_j may be found at each stage j such that (1.23a) has order $m-j$, and we only have to prove that (1.23) defines approximants which interpolate correctly at $z_0, z_1, \ldots, z_m$ to the values given by (1.21). From (1.23), it follows that

$$D = \begin{vmatrix} p^{[m-j/j]}(z) & p^{[m-j+1/j-1]}(z) \\ q^{[m-j/j]}(z) & q^{[m-j+1/j-1]}(z) \end{vmatrix}$$

$$= \begin{vmatrix} p^{[m-j+1/j-1]}(z) & p^{[m-j+2/j-2]}(z) \\ q^{[m-j+1/j-1]}(z) & q^{[m-j+2/j-2]}(z) \end{vmatrix}.$$

D is therefore independent of j, and (1.21), (1.22) imply that

$$D = -\prod_{k=0}^{m} (z - z_k).$$

Thus, provided the algorithm is nonsingular,

$$\frac{p^{[m-j/j]}(z)}{q^{[m-j/j]}(z)} = \frac{p^{[m-j+1/j-1]}(z)}{q^{[m-j+1/j-1]}(z)} - \frac{\prod_{k=0}^{m} (z - z_k)}{q^{[m-j/j]}(z) q^{[m-j+1/j-1]}(z)},$$

showing that each entry interpolates as required. Notice that this algorithm is unaffected by confluence once it has been initialized.

An attractive feature of the Kronecker algorithm is that it is easily modified to become a reliable algorithm. If, for some value of j, $p^{[m-j/j]}(z)$ defined by (1.23a) does not have its full indicated order $(m-j)$, the $[m-j/j]$ interpolant is degenerate. The next interpolant in the sequence is defined by

$$p^{[m-j-k/j+k]}(z) = \pi_k(z) p^{[m-j/j]}(z) - p^{[m-j+1/j-1]}(z), \quad (1.23c)$$

$$q^{[m-j-k/j+k]}(z) = \pi_k(z) q^{[m-j/j]}(z) - q^{[m-j+1/j-1]}(z), \quad (1.23d)$$

where $\pi_k(z)$ is the polynomial of least order k such that the left-hand side of (1.23c) has its indicated order $(m-j-k)$. It is easy to show [Graves-Morris, 1979], that (1.23c) and (1.23d) define the next nondegenerate interpolant in the antidiagonal sequence. Kronecker's algorithm would appear to be especially useful for the case where exact arithmetic is available and any degeneracy test must be decisive.

THIELE'S RECIPROCAL DIFFERENCE METHOD [Thiele, 1909, Chapter 3; Hildebrand, 1956]. This method yields a continued-fraction representation of the N-point Padé approximant. Normally, the interpolation points should be distinct, although the elements of the fraction may be defined by continuity in the presence of confluence. Define the reciprocal differences

$$\rho_0 \equiv \rho[z_0] = f(z_0),$$ (1.24a)

$$\rho_1 \equiv \rho[z_0, z_1] = (z_0 - z_1)\{f(z_0) - f(z_1)\}^{-1},$$ (1.24b)

$$\rho[z, z_0] = (z - z_0)\{f(z) - f(z_0)\}^{-1},$$ (1.24c)

and in general for $n > 1$,

$$\rho_n \equiv \rho[z_0, z_1, \ldots, z_{n-1}, z_n],$$ (1.24d)

$$\rho[z_0, z_1, \ldots z_{n-1}, z_n] = \frac{z_0 - z_n}{\rho[z_0, \ldots, z_{n-1}] - \rho[z_1, \ldots, z_{n-1}, z_n]} + \rho[z_1, \ldots, z_{n-1}].$$ (1.24e)

Then the interpolant defined on $z_0, z_1, \ldots, z_n$ is given by

$$r_n(z) = \rho_0 + \frac{z - z_0}{\rho_1} + \frac{z - z_1}{\rho_2 - \rho_0} + \frac{z - z_2}{\rho_3 - \rho_1} + \cdots + \frac{z - z_{n-1}}{\rho_n - \rho_{n-2}}.$$ (1.25)

Notice that if $n = 2M$, which means that an odd number of interpolation points are used, then (1.25) defines an $[M/M]$-type approximant; if $n = 2M + 1$, an even number of points are used and an $[M+1/M]$-type approximant is defined.

Verification. We first prove by induction the identity

$$f(z) = \rho_0 + \frac{z - z_0}{\rho_1} + \frac{z - z_1}{\rho_2 - \rho_0} + \frac{z - z_2}{\rho_3 - \rho_1} + \cdots + \frac{z - z_n}{\rho[z, z_0, \ldots, z_n] - \rho_{n-1}}$$ (1.26)

For $n = 0$, (1.26) is interpreted as

$$f(z) = \rho_0 + \frac{z - z_0}{\rho[z, z_0]},$$

which is verified by (1.24c). For $n > 0$, we expand the final denominator of (1.26) using the identity

$$\rho[z, z_0, \ldots, z_n] - \rho_{n-1} = \rho_{n+1} - \rho_{n-1} + \frac{z - z_{n+1}}{\rho[z, z_0, \ldots, z_{n+1}] - \rho_n},$$

which is derived from (1.24e) in the form

$$\rho[z, z_0, \ldots, z_{n+1}] = \frac{z - z_{n+1}}{\rho[z, z_0, \ldots, z_n] - \rho_{n+1}} + \rho_n.$$

Hence (1.26) is proved by induction. By letting $z=z_0, z_1,\ldots, z_n$ sequentially, and provided no unusual cancellation occurs (1.26) shows that (1.25) interpolates correctly at the $n+1$ interpolation points, and therefore (1.25) is the $(n+1)$-point Padé approximant.

Thiele's method of rational interpolation is more interesting from an analytical point of view; for numerical values the following computational scheme is as efficient as any.

MODIFIED THACHER–TUKEY ALGORITHM [Thacher and Tukey, 1960; Graves-Morris and Hopkins, 1981]. We reexpress (1.25) as

$$r_n(z)=f(z_0)+\frac{a_1(z-z_0)}{1}+\frac{a_2(z-z_1')}{1}+\frac{a_3(z-z_2')}{1}+\cdots+\frac{a_n(z-z_{n-1}')}{1}.$$

(1.27)

The interpolation set is the set of distinct points

$$S_0=\{z_0, z_1,\ldots, z_n\},$$

which are used in the order $z_0, z_1', z_2',\ldots, z_n'$.

To motivate the algorithm, we consider a function $f(z)$ defined on the interpolation set S_0 and interpolated by (1.27). We define functions $g_0(z), g_1(z),\ldots, g_n(z)$ by

$$f(z)=f(z_0)+g_0(z),$$

(1.28a)

and iteratively by

$$g_{i-1}(z)=\frac{a_i(z-z_{i-1}')}{1+g_i(z)} \qquad \text{for} \quad i=0,1,2,\ldots, n.$$

(1.28b)

The special case of $f(z)=r_n(z)$ corresponds to $g_n(z)\equiv 0$. We expect, from (1.28b), that $g_{i-1}(z_{i-1}')=0$, and hence also from this equation we expect that

$$g_{i-1}(z_i')=a_i(z_i'-z_{i-1}').$$

This equation is re-expressed in (1.29) to evaluate the coefficients $a_1, a_2,\ldots, a_n$, which must be finite and nonzero. It is part of the normal iterative step of the algorithm, called stage (a). If it turns out at some juncture (with $j=t+1$) that $g_t(z)=0$ for all $z\in S_{t+1}$, where S_{t+1} is a residual interpolation set, then this corresponds to stage (b) of the algorithm in which $r_t(z)$ is apparently the correct but degenerate interpolant. Certainly it is true that in all other cases, corresponding to stage (c) of the algorithm, there is no interpolating rational function. Having constructed (1.27) as the proposed interpolant, one must check that its denominator $q_t(z)$ given by (1.30) does not vanish at any of the interpolation points. One may show that this circumstance corresponds uniquely to the nonexistence of an interpolant. Notice that this algorithm is reliable in the sense that if an interpolant to the data exists, the algorithm finds it; if no such interpolant exists, the algorithm detects this situation via an error exit.

Initialization.   We define the set

$$S_1 = \{z_1, z_2, \ldots, z_n\}.$$

We define values of a function $g_0(z)$ for $z \in S_1$ by

$$g_0(z) = f(z) - f(z_0) \qquad \text{for} \quad z \in S_1.$$

Iteration.   The normal iterative step, used for $j = 1, 2, \ldots$ until termination, begins with stage (a). Otherwise, the iteration is concluded by stage (b) or stage (c).

Stage (a). If possible, choose z_j' from S_j such that

$$g_{j-1}(z_j') \neq 0, \infty.$$

Then we define

$$\left.\begin{aligned}
a_j &= \frac{g_{j-1}(z_j')}{z_j' - z_{j-1}'}, \\[2mm]
S_{j+1} &= S_j \setminus z_j', \\[2mm]
g_j(z) &= \frac{a_j(z - z_{j-1}')}{g_{j-1}(z)} - 1 \qquad \text{for all} \quad z \in S_{j+1}.
\end{aligned}\right\} \quad (1.29)$$

If $j = n$, we set $t = n$ and proceed to the termination stage; otherwise we proceed with the iteration with $j := j + 1$.

If it is not possible to choose z_j' from S_j such that

$$g_{j-1}(z_j') \neq 0, \infty,$$

it may be that we are at stage (b).

Stage (b). This stage is reached if, and only if,

$$g_{j-1}(z) = 0 \qquad \text{for all} \quad z \in S_j.$$

In this case, set $t = j - 1$ in (1.29). Exit from the iteration and proceed to termination for the denominator check.

Stage (c). This stage is reached if, and only if $g_{j-1}(z)=0, \infty$ for all $z \in S_j$, but $g_{j-1}(z) \neq 0$ for some $z \in S_j$. The algorithm terminates prematurely with an error exit, signifying that the desired interpolant does not exist.

Termination. If termination is reached with $t=0$, then $r(z)=f_0$ is the correct rational interpolant. If termination is reached with $t=1,2,\ldots,n$, we define

$$q_1(z)=1, \qquad q_2(z)=1+a_2(z-z_1')$$ (1.30a)

and for $i=2,3,\ldots,t-1$,

$$q_{i+1}(z)=q_i(z)+a_{i+1}(z-z_i')q_{i-1}(z).$$ (1.30b)

Provided $q_t(z) \neq 0$ for all $z \in S_0$, the algorithm has a successful exit. Otherwise $q_t(z_j)=0$ for some j, $0 \leqslant j \leqslant n$, and the algorithm has a failure exit in the termination stage, signifying that the required interpolant does not exist.

GENERALIZED Q.D. ALGORITHM [Wuytack, 1973; Graves-Morris, 1980b]. The Q.D. algorithm may be used to derive a continued-fraction representation of a Padé approximant from the coefficients of a Maclaurin series (see Part I, Section 4.4). We consider a generalization of this algorithm to the case in which the interpolation points may be distinct. We consider the following Thiele-type interpolant:

$$g_0(z)=\frac{c_0}{1} - \frac{q_1^0(z-z_0)}{1} - \frac{e_1^0(z-z_1)}{1} - \frac{q_2^0(z-z_2)}{1} - \frac{e_2^0(z-z_3)}{1} \cdots .$$

(1.31)

The coefficients of the nth convergent of $g_0(z)$ are derived from the coefficients of the Newton interpolating polynomial

$$\pi_n(z)=c_0+c_1(z-z_0)+c_2(z-z_0)(z-z_1)+ \cdots +c_n \prod_{i=0}^{n-1}(z-z_i)$$

using the following algorithm:

Initialization. For $J=0,1,2,\ldots,n-1$, define

$$Z_1^J=z_{J+1}-z_J,$$ (1.32a)

$$e_0^{J+1}=0,$$ (1.32b)

$$q_1^J=\left[Z_1^J+\frac{c_J}{c_{J+1}}\right]^{-1},$$ (1.32c)

$$e_1^J=-q_1^J-q_1^{J+1}(q_1^J Z_1^J-1).$$ (1.32d)

Recurrence. For $J=0,1,2,\ldots$ and $i=2,3,\ldots$, we construct all well-defined quantities q_i^J, e_i^J recursively from the formulas

$$Z_i^J = z_{J+2i-1} - z_{J+2i-2},\tag{1.33a}$$

$$q_i^J = \left[Z_i^J - \frac{e_{i-1}^J}{e_{i-1}^J + q_{i-1}^J} \frac{q_{i-1}^{J+1} + e_{i-2}^{J+1}}{q_{i-1}^{J+1}} \frac{Z_i^J e_{i-1}^{J+1} - 1}{e_{i-1}^{J+1}} \right]^{-1},\tag{1.33b}$$

$$e_i^J = -q_i^J + \left(Z_i^J q_i^J - 1 \right) \left(e_{i-1}^{J+1} + q_i^{J+1} \right) \left(Z_i^J e_{i-1}^{J+1} - 1 \right)^{-1}.\tag{1.33c}$$

Notice that this algorithm is well defined in the presence of confluence, in which case it reduces to a minor variant of the Q.D. algorithm (I.4.4.31, 32).

GENERALIZED ε-ALGORITHM [Claessens, 1978c]. This algorithm is suitable for calculating the value of a rational interpolant. It is based on Claessens's identity, which, in the notation of (1.12), (1.13), and (1.14), is

$$\left[\left\{ r^{[L+1/M]}(z) - r^{[L/M]}(z) \right\}^{-1} - \left\{ r^{[L/M+1]}(z) - r^{[L/M]}(z) \right\}^{-1} \right] (z - z_{L+M})$$

$$= \left[\left\{ r^{[L/M-1]}(z) - r^{[L/M]}(z) \right\}^{-1} \right.$$

$$\left. - \left\{ r^{[L-1/M]}(z) - r^{[L/M]}(z) \right\}^{-1} \right] (z - z_{L+M+1})\tag{1.34}$$

whenever the indicated quantities exist and are nondegenerate. Claessens's identity reduces to Wynn's identity in the confluent limit.

Outline proof of (1.34). We follow the method of Part I, Section 3.4. We define

$$F_{1,L+M}^{[L/M]} = \begin{vmatrix} f_{M,L+1} & f_{M,L+2} & \cdots & f_{M,L+M} \\ f_{M-1,L+1} & f_{M-1,L+2} & \cdots & f_{M-1,L+M} \\ \vdots & \vdots & & \vdots \\ f_{1,L+1} & f_{1,L+2} & \cdots & f_{1,L+M} \end{vmatrix}\tag{1.35}$$

and note that $F_{1,L+M}^{[L/M]} = C(L/M)$ in the confluent limit. The subscripts of $F_{1,L+M}^{[L/M]}$ denote the indices $1,2,\ldots,L+M$ of the interpolation points used in its construction. Using the methods of Part I, Section 3.4, we find that (I.3.4.4) generalizes directly to

$$r^{[L+1/M]}(z) - r^{[L/M]}(z) = \frac{(z - z_0) \cdots (z - z_{L+M+1}) F_{0,L+M+1}^{[L+1/M+1]} F_{0,L+M}^{[L+1/M]}}{v^{[L+1/M]}(z) v^{[L/M]}(z)},$$

$$\tag{1.36}$$

and (I.3.4.6) generalizes directly to

$$r^{[L/M+1]}(z)-r^{[L/M]}(z)=\frac{(z-z_0)\cdots(z-z_{L+M+1})F_{0,L+M}^{[L/M+1]}F_{0,L+M+1}^{[L+1/M+1]}}{v^{[L/M+1]}(z)v^{[L/M]}(z)}.$$

$$(1.37)$$

By reordering the points of (1.18), we find that

$$v^{[L/M+1]}(z)$$

$$=\begin{vmatrix} f_{M,L} & f_{M-1,L} & \cdots & f_{0,L} & \prod_{k=L+1}^{L+M+1}(z-z_k) \\ f_{M,L+1} & f_{M-1,L+1} & \cdots & f_{0,L+1} & \prod_{k=L+2}^{L+M+1}(z-z_k) \\ \vdots & \vdots & & \vdots & \vdots \\ f_{M,L+M+1} & f_{M-1,L+M+1} & \cdots & f_{0,L+M+1} & 1 \end{vmatrix}.$$

By applying Sylvester's identity to this we find that

$$v^{[L/M+1]}(z)F_{0,L+M}^{[L+1/M]}=v^{[L+1/M]}(z)F_{0,L+M}^{[L/M+1]}$$

$$-(z-z_{L+M+1})F_{0,L+M+1}^{[L+1/M+1]}v^{[L/M]}(z),\quad (1.38)$$

which is a generalization of (I.3.4.2).
  Hence we deduce from (1.36), (1.37), and (1.38) that

$$\{r^{[L+1/M]}(z)-r^{[L/M]}(z)\}^{-1}-\{r^{[L/M+1]}(z)-r^{[L/M]}(z)\}^{-1}$$

$$=\frac{\{v^{[L/M]}(z)\}^2(z-z_0)^{-1}\cdots(z-z_{L+M})^{-1}}{F_{0,L+M}^{[L+1/M]}F_{0,L+M}^{[L/M+1]}},$$

which is a generalization of (I.3.4.8). Equation (1.34) follows from a similar treatment of the right-hand side.
  The generalized ε-algorithm is the formal identity

$$(z-z_{k+j+1})[\varepsilon_{k+1}^{(j)}-\varepsilon_{k-1}^{(j+1)}][\varepsilon_k^{(j+1)}-\varepsilon_k^{(j)}]=1\qquad (1.39a)$$

for indices k,j in the range $k=0,1,2,\ldots$ and $j\geq-[k/2]$. The artificial initialization conditions are

$$\varepsilon_{-1}^{(j)}=0,\qquad j=0,1,2,\ldots,$$

and

$$\varepsilon_{2k}^{(-k-1)} = 0, \qquad k = 0, 1, 2, \ldots . \tag{1.39b}$$

The usual initialization condition, using values derived from an interpolating polynomial, is

$$\varepsilon_0^{(j)} = r^{[j/0]}(z). \tag{1.39c}$$

Elements of the ε-table (see Part I, Section 3.3, Table 1, extended above the diagonal in Part I, Section 3.6, Table 2) are identified with values of rational interpolants by the formula

$$\varepsilon_{2k}^{(j)} = r^{[k+j/k]}(z), \qquad k = 0, 1, 2, \ldots, \quad j \geqslant -k. \tag{1.39d}$$

Proof of the results (1.39a)–(1.39d) is based on (1.34) and closely follows the corresponding argument of Part I, Section 3.4.

The foregoing summary of methods of rational interpolation is biased toward methods which reduce to Padé methods in the confluent limit. Possibly the neatest of the continued-fraction algorithms for rational interpolation at distinct points is the t-g algorithm of Claessens [1976]. A recent algorithm of Werner [1979], which leads to a Thiele–Werner interpolating rational fraction, has many advantages. It is based on principles similar to those of the generalized Viskovatov algorithm (see Part I, Section 4.5); it is a reliable algorithm, it has a confluent limit, and it is easily adapted to ensure numerical stability [Graves-Morris, 1980a]. The original papers on rational interpolation [Cauchy, 1821; Jacobi, 1846; Thacher and Tukey, 1960; Stoer, 1961; Wetterling, 1963; Larkin, 1967] lead to the methods developed in the review of Werner and Schaback [1972].

The problems of normality and degeneracy are reviewed by Meinguet [1970], and we refer to Gallucci and Jones [1976] and Claessens [1978a, b] for details of normality and degeneracy in the context of the Newton–Padé table. We refer to Walsh [1964a, b, 1965a, b], Saff [1972], Karlsson [1976], and Gončar and Guillermo López [1978] for discussions of rational interpolation within the context of rational approximation. Bounding properties of rational interpolants to Stieltjes functions are given by Baker [1969] and Barnsley [1973].

By way of applications, we mention two important areas out of the many applications of rational interpolation. There has always been considerable interest in Padé-type methods as algorithms for finding zeros of functions. There is no difficulty in finding high-order, interpolatory methods [Merz, 1968; Zinn-Justin, 1970; Larkin, 1981], but in this context the principal difficulties are connected with minimizing the number of operations associated with the slowest path the algorithm may take, and avoidance of branching decisions based on noise [Garside et al., 1968; Dekker, 1969; Jarratt, 1970; Bus and Dekker, 1975].

Rational interpolation is a popular means of deferred extrapolation to the limit $h \rightarrow 0$ in the context of the numerical solution of ordinary differential equations using a finite grid of spacing h. This technique was pioneered by Bulirsch and Stoer [1964] and Gragg [1965]. Rational forms of the solution are used by Lambert and Shaw [1965, 1966] and Luke et al. [1975].

This completes our summary of the principal algorithms for N-point Padé approximation. A simple application of N-point Padé approximation to the acceleration of convergence is given in Section 1.3 Next, we turn our attention to the special case of $N=2$, as described by Equations (1.1), (1.2), and (1.3). These have been extensively investigated within the framework of continued fractions, and are often called general T-fractions after Thron [1948].

Let us assume that we are given a pair of formal power series for a function $f(z)$:

$$\mathcal{L} = \sum_{i=0}^{\infty} c_i z^i = c_0 + c_1 z + c_2 z^2 + \cdots, \tag{1.40a}$$

$$\tilde{\mathcal{L}} = \sum_{i=0}^{\infty} \tilde{c}_{1-i} z^{1-i} = \tilde{c}_1 z + \tilde{c}_0 + \tilde{c}_{-1} z^{-1} + \tilde{c}_{-2} z^{-2} + \cdots, \tag{1.40b}$$

where $\tilde{\mathcal{L}}$ given by (1.40b) is a Laurent series of $f(z)$ about $z = \infty$ having a finite principal part. The representations (1.40) normally define the coefficients $\{e_i, d_i, i = 0, 1, 2, \dots\}$ of the general T-fraction

$$T(z) = e_0 + d_0 z + \cfrac{z}{e_1 + d_1 z} + \cfrac{z}{e_2 + d_2 z} + \cdots, \tag{1.41}$$

in which $e_i \neq 0$, $d_i \neq 0$ for $i = 1, 2, 3, \dots$. We refer to Part I, Section 4.5, Exercise 2 for an example.

The Jones–Thron theorem demonstrates what has been achieved so far using general T-fractions.

THEOREM 1.1.2 [Jones, 1977]. *Suppose a pair of formal Laurent series*

$$\mathcal{L} = \sum_{k=-\nu}^{\infty} c_k z^k \quad and \quad \tilde{\mathcal{L}} = \sum_{k=-\infty}^{\mu} \tilde{c}_k z^k$$

with $\nu \geq 0$, $\mu \geq 0$ are given. Then there exists a general T-fraction (generalizing (1.30), (1.31))

$$T(z) = \sum_{k=1}^{\mu} \tilde{c}_k z^k + \sum_{k=-\nu}^{0} c_k z^k + \cfrac{F_1 z}{1 + G_1 z} + \cfrac{F_2 z}{1 + G_2 z} + \cdots \tag{1.42}$$

with $F_n \neq 0$ and $G_n \neq 0$ such that the nth convergent of $T(z)$ has Laurent expansions of the forms

$$T_n(z) = c_{-\nu} z^{-\nu} + c_{-\nu+1} z^{-\nu+1} + \cdots + c_n z^n + \cdots \qquad at \quad z = 0,$$

$$T_n(z) = \tilde{c}_\mu z^\mu + \tilde{c}_{\mu-1} z^{\mu-1} + \cdots + \tilde{c}_n z^{-n+1} + \cdots \qquad at \quad z = \infty,$$

provided the determinants $\Delta_n \neq 0$ and $\Phi_n \neq 0$ for $n = 1, 2, 3, \ldots$. These determinants are defined by letting $c_k = 0$ for $k < -\nu$, $\tilde{c}_k = 0$ for $k > \mu$, $\delta_k = \tilde{c}_k - c_k$ for all k, $\Delta_{-1} = \Delta_0 = \Phi_0 = -\Phi_{-1} = 1$, and

$$\Delta_n = \begin{vmatrix} \delta_{1-n} & \cdots & \delta_0 \\ \vdots & & \vdots \\ \delta_0 & \cdots & \delta_{n-1} \end{vmatrix}, \qquad \Phi_n = \begin{vmatrix} \delta_{2-n} & \cdots & \delta_1 \\ \vdots & & \vdots \\ \delta_1 & \cdots & \delta_n \end{vmatrix}$$

for $n \geq 1$. Moreover, the coefficients in the general T-fraction (1.32) are given by

$$F_1 = -\Phi_1, \qquad F_n = -\frac{\Delta_{n-1} \Phi_{n-1}}{\Delta_{n-2} \Phi_n}, \quad n \geq 2,$$

and

$$G_n = -\frac{\Delta_{n-1} \Phi_n}{\Delta_n \Phi_{n-1}}, \qquad n \geq 1.$$

We refer to the companion volume of Jones and Thron for a proof of this theorem. Essentially, this theorem specifies necessary conditions for the existence of a class of two-point Padé approximants. However, it does not answer the convergence question at all. If the formal Laurent series $\mathcal{L}$ and $\tilde{\mathcal{L}}$ in (1.40) represent one given function, one expects the general T-fractions to be efficient approximations subject to certain constraints. But if the series $\mathcal{L}$ and $\tilde{\mathcal{L}}$ represent different functions, construction of the "associated" T-fraction should prove meaningless. This is borne out by Thron's example:

$$T(z) = 1 - z + \frac{z}{1-z} + \frac{z}{1-z} + \cdots.$$

Using the recurrence relations (I.4.4.4), we find that (see Part I, Section 4.4, Exercise 1)

$$T(z) = 1 \qquad \text{for } |z| < 1,$$
$$T(z) = -z \qquad \text{for } |z| > 1.$$

All the poles and zeros of the convergents lie on the unit circle. This summarizes what occurs when the T-fraction method is applied to the unnatural problem of simultaneously fitting the series $\ell = 1$ and $\tilde{\ell} = -z$.

The next result [Dijkstra, 1977; Wynn, 1962a] is a good example of a T-fraction expansion. It is

$$\frac{{}_1F_1(a,b+1;z)}{{}_1F_1(a,b;z)} = \frac{b}{b+z} - \frac{z(b+1-a)}{b+1+z} - \cdots - \frac{z(b+n-a)}{b+n+z} - \cdots$$

with $a, b \geqslant 0$ and $z \geqslant 0$. It embodies the Maclaurin expansion about $z = 0$ and the asymptotic expansion about $z = +\infty$ of the left-hand side. The special case with $b = a$ is

$$\frac{{}_1F_1(a,a+1;z)}{e^z} = \frac{a}{a+z} - \frac{z}{a+1+z} - \cdots - \frac{nz}{a+n+z} - \cdots .$$

Using Kummer's transformation, this becomes

$$ {}_1F_1(1,a+1;z) = \frac{a}{a-z} + \frac{a}{a+1+z} + \cdots + \frac{nz}{a+n+z} + \cdots . \quad (1.43)$$

The original restriction to $z \geqslant 0$ may now be dropped, because the right-hand side of (1.43) is convergent throughout the z-plane [Jones et al., 1979]. This formula is important because it leads to a T-fraction development of the error function, the incomplete gamma function, and the generalized Dawson integral. We define [Dijkstra, 1977]

$$F(p,x) = e^{-x^p} \int_0^x e^{t^p}\, dt, \qquad x \geqslant 0, \quad p > 0,$$

to be the generalized Dawson integral; Dawson's integral is $D(x) = F(2, x)$, and its relation to the error function is expressed by

$$\mathrm{erf}(z) = \frac{2i}{\sqrt{\pi}} e^{-z^2} D(-iz). \qquad (1.44)$$

Then we find (see Part I, Section 4.6)

$$\gamma(a,z) = z^a e^{-z} a^{-1} {}_1F_1(1,1+a;z), \qquad\qquad (1.45a)$$

$$F(p,x) = x\, {}_1F_1(1,a+1;-z), \qquad \text{with} \quad z = x^p \text{ and } a = p^{-1}, \quad (1.45b)$$

$$D(z) = z\, {}_1F_1(1,\tfrac{3}{2};-z^2), \qquad\qquad (1.45c)$$

and

$$\operatorname{erf}(z) = \frac{2ze^{-z^2}}{\pi} {}_1F_1\left(1, \tfrac{3}{2}; z^2\right). \tag{1.45d}$$

T-fraction developments of all these functions (1.45a–d) then follow from (1.43) and are valid for all z.

For further details on the subject of T-fractions, we refer to Murphy [1971], McCabe [1974, 1975], McCabe and Murphy [1976], Drew and Murphy [1977], Waadeland [1979], Sidi [1980a] and Jones, Thron, and Waadeland [1980].

Exercise. Prove that F_j, G_j in (1.42) are generated by the following Q.D. algorithm [McCabe, 1975]: The Q.D. algorithm for T-fractions is initialized by

$$F_1^{(k)} = 0, \quad G_1^{(k)} = -\frac{\delta_{k+1}}{\delta_k}, \quad k = 0, 1, -1, 2, -2, \ldots.$$

The elements are defined recursively for $k = 0, 1, -1, 2, -2, \ldots$ and $j = 1, 2, 3, \ldots$ by rhombus rules:

$$F_{j+1}^{(k)} + G_j^{(k)} = F_j^{(k+1)} + G_j^{(k+1)} \qquad \begin{pmatrix} & * & * \\ * & * & \end{pmatrix}$$

and

$$F_{j+1}^{(k)} \cdot G_{j+1}^{(k+1)} = F_{j+1}^{(k+1)} \cdot G_j^{(k)} \qquad \begin{pmatrix} * & * & \\ & * & * \end{pmatrix}.$$

The quantities F_i, G_i in (1.42) are given by

$$F_1 = c_1, \quad G_1 = G_1^{(0)}$$

and for $j = 2, 3, 4, \ldots$

$$F_j = F_j^{(0)}, \quad G_j = G_j^{(0)}.$$

In this case, the Q.D. table consists of entries arranged as

$\vdots$	$\vdots$	$\vdots$	$\vdots$	$\vdots$	
$F_1^{(-1)}$	$G_1^{(-1)}$	$F_2^{(-1)}$	$G_2^{(-1)}$	$F_3^{(-1)}$	$\cdots$
$F_1^{(0)}$	$G_1^{(0)}$	$F_2^{(0)}$	$G_2^{(0)}$	$F_3^{(0)}$	$\cdots \leftarrow$ principal row
$F_1^{(1)}$	$G_1^{(1)}$	$F_2^{(1)}$	$G_2^{(1)}$	$F_3^{(1)}$	$\cdots$
$\vdots$	$\vdots$	$\vdots$	$\vdots$	$\vdots$	

The first two columns are defined by the initialization, and the Q.D. algorithm determines the elements of the principal row which are needed in (1.42).

1.2 Baker–Gammel Approximants

A Padé–Legendre series is a series

$$g(z) = \sum_{n=0}^{\infty} c_n P_n(z) = \sum_{n=0}^{\infty} f_n P_n(-z) \tag{2.1}$$

where $\{P_n(z),\ n=0,1,2,\ldots\}$ are Legendre polynomials and $f_n=(-)^n c_n$. In Section 1.6 we discuss rational approximants of $g(z)$, whereas in this section we discuss an alternative class of approximating functions suitable for (2.1) and similar series. Except in degenerate cases the coefficients c_n are expressed as

$$c_n = \sum_{i=1}^{M} \alpha_i h_i^n \qquad \text{for} \quad n=0,1,2,\ldots,2M-1. \tag{2.2}$$

If we suppose that (2.2) is approximately true for $n=2M,2M+1,\ldots$, which is to say that we suppose that the coefficients c_n behave in a generalized geometric manner, then we see that $g(z)$ is approximated by

$$G^{[M-1/M]}(z) = \sum_{i=1}^{M} \alpha_i \left(1 - 2h_i z + h_i^2\right)^{-1/2}. \tag{2.3}$$

Equation (2.3) is based on a generating function for the Legendre polynomials and provides a nonrational approximation for $g(z)$. To determine the parameters $\{\alpha_i, h_i,\ i=1,2,\ldots,M\}$ occurring in (2.3), we deduce from (2.2) that

$$f(z) \equiv \sum_{n=0}^{\infty} c_n z^n = \sum_{i=1}^{M} \frac{\alpha_i}{1 - h_i z} + O(z^{2M}) \tag{2.4}$$

is formally valid, so that h_i^{-1} are the poles of the $[M-1/M]$ Padé approximant of $f(z)$ and $-\alpha_i/h_i$ are the corresponding residues. Thus the approximation (2.3) is easily constructed provided that the $[M-1/M]$ Padé approximant of $f(z)$ defined by (2.3) exists.

The ideas underlying the approach expressed by (2.1)–(2.4) evolved from the papers of Gammel et al. [1967], Baker [1967], and Common [1969a, b]. A rather different approach, based on the theory of positive functionals and different hypotheses from those of this section, leading to the inequalities of

Theorem 1.2.2, was derived by M. Riesz [1923]; Shohat extended this approach to the case of Hamburger series [Akhiezer, 1965]. Some of the bounds obtained in this section may be compared with those obtained by the methods of Section 3.2.

Baker–Gammel approximants exploit and extend these ideas by considering functions having a formal expansion

$$g(z) = \sum_{m=0}^{\infty} f_m k_m(z)$$

which generalizes (2.1), and the integral representation

$$g(z) = \int_0^{\infty} k(z, u) \, d\phi(u), \tag{2.5}$$

where $k(z, u)$ is a kernel, yet to be specified, and $d\phi(u)$ is a Stieltjes measure.

For the special case that $k(z, u) = (1 + uz)^{-1}$, (2.5) becomes a Stieltjes series, and the methods of Part I, Chapter 5 should be used. More generally, we begin by constructing Padé approximants to $f(z)$ defined by

$$f(z) \equiv \sum_{n=0}^{\infty} c_n z^n \equiv \int_0^{\infty} \frac{d\phi(u)}{1 + zu} = [M + J/M] + O(z^{2M+J+1}), \tag{2.6}$$

which is a Stieltjes series. Baker–Gammel approximants for $g(z)$ in (2.5) are defined by

$$G^{[M+J/M]}(z) = \sum_{j=0}^{J} \beta_j k_j(z) + \sum_{i=1}^{M} \alpha_i k(z, u_i), \tag{2.7}$$

where $k(z, u)$ is defined so that

$$k_j(z) = \frac{1}{j!} \left(\frac{\partial}{\partial u} \right)^j k(z, u) \Bigg|_{u=0}, \qquad j = 0, 1, \dots. \tag{2.8}$$

The values of β_j, $j = 0, 1, \dots, J$, and u_i, α_i, $i = 1, 2, \dots, M$, remain to be specified; if $J = -1$ the polynomial term of (2.7) is absent.

Ignoring, for the moment, all questions of convergence in (2.9) and (2.10), (2.8) leads to

$$k(z, u) = \sum_{m=0}^{\infty} u^m k_m(z). \tag{2.9}$$

Then (2.5), (2.6), and (2.9) show that $g(z)$ has the series expansion

$$g(z) = \sum_{m=0}^{\infty} f_m k_m(z),\tag{2.10}$$

where each f_m is represented by

$$f_m = (-)^m c_m = \int_0^{\infty} u^m \, d\phi(u), \qquad m = 0, 1, \dots.\tag{2.11}$$

From (2.10), (2.7), and (2.9), we see that the coefficients of $k_m(z)$ in $g(z)$ and $G^{[M+J/M]}(z)$ are equal for $m = 0, 1, \dots, 2M+J$ provided

$$f_m = \beta_m + \sum_{i=1}^{M} \alpha_i u_i^m, \qquad m = 0, 1, \dots, J,\tag{2.12}$$

and

$$f_m = \sum_{i=1}^{M} \alpha_i u_i^m, \qquad m = J+1, J+2, \dots, 2M+J.\tag{2.13}$$

From (2.12) and (2.13), we find formally that

$$\sum_{m=0}^{\infty} f_m(-z)^m = \sum_{j=0}^{J} \beta_j(-z)^j + \sum_{i=1}^{M} \frac{\alpha_i}{1+zu_i} + O(z^{2M+J+1}).\tag{2.14}$$

Hence we deduce that the parameters β_j, α_i, u_i occurring in (2.7) are defined by (2.6) with the identification

$$[M+J/M]_f(z) = \sum_{j=0}^{J} \beta_j(-z)^j + \sum_{i=1}^{M} \frac{\alpha_i}{1+zu_i}.\tag{2.15}$$

This establishes that existence of the $[M+J/M]$ Padé approximant for $f(z)$ defined by (2.6) is sufficient formally to derive a Baker–Gammel approximant for $g(z)$ defined by (2.5). The justification of the former procedure lies in the following theorem.

THEOREM 1.2.1. *Let $k(z, u)$ be analytic on the positive real u-axis and at least within a distance Δ from it for all $z \in \mathcal{C}$, a compact region in the z-plane. Further, let $|k(z, u)|$ be uniformly bounded by $[\ln u]^{-(1+\mu)}$ with $\mu > 0$ as $u \to +\infty$. Then*

$$G^{[M+J/M]}(z) \to g(z) \quad \text{for } z \in \mathcal{C}, \qquad \text{as } M \to \infty.$$

Proof. With a Cauchy representation for $k(z, u)$, (2.5) becomes

$$g(z) = \int_0^\infty \frac{d\phi(u)}{2\pi i} \int_{\Gamma_1} \frac{k(z, w)}{w - u} dw,$$

where Γ_1 is a circular contour with center u and radius Δ. Using the analyticity properties of $k(z, w)$ assumed, we deduce that

$$g(z) = \frac{1}{2\pi i} \int_0^\infty d\phi(u) \int_\Gamma \frac{k(z, w)}{w - u} dw,$$

where Γ is a contour at distance Δ from the positive real w-axis, as shown in Figure 1.

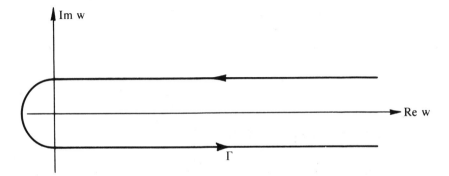

Figure 1. The contour Γ in the complex w-plane.

Hence,

$$g(z) = \frac{1}{2\pi i} \int_\Gamma \frac{k(z, w)}{w} dw \int_0^\infty \frac{d\phi(u)}{1 - u/w}$$

$$= \frac{1}{2\pi i} \int_\Gamma k(z, w) f(-1/w) \frac{dw}{w}. \tag{2.16}$$

Baker–Gammel approximants are formed by replacing the function f in (2.16) with its Padé approximant (2.15), so that

$$G^{[M+J/M]}(z) = \frac{1}{2\pi i} \int_\Gamma k(z, w) \left[\sum_{j=0}^J \beta_j w^{-j} + \sum_{i=1}^M \frac{\alpha_i}{1 - u_i/w} \right] \frac{dw}{w}. \tag{2.17}$$

Because $f(z)$ is a Stieltjes series, the poles at $w = u_i$ in (2.17) lie inside Γ, and we deduce from (2.8) that

$$G^{[M+J/M]}(z) = \sum_{j=0}^J \beta_j k_j(z) + \sum_{i=1}^M \alpha_i k(z, u_i).$$

Thus the heuristic arguments connecting (2.6), (2.7), and (2.15) are rigorously justified under the conditions of the theorem. The error formula, which follows from (2.16) and (2.17), is

$$g(z) - G^{[M+J/M]}(z)$$

$$= \frac{1}{2\pi i} \int_\Gamma \frac{k(z,w)}{w} \left\{ f\left(\frac{-1}{w}\right) - \sum_{j=0}^{J} \beta_j w^{-j} - \sum_{i=1}^{M} \frac{\alpha_i}{1 - u_i/w} \right\} dw. \quad (2.18)$$

Using Theorems I.5.2.6 and I.5.3.1, we deduce that the term in brackets in (2.18) tends uniformly to zero as $M \to \infty$ for $w \in \Gamma$; furthermore it is $O(w^{-1})$ as $w \to \infty$. Hence (2.18) tends to zero as $M \to \infty$ for $z \in \mathcal{C}$, and convergence of the Baker–Gammel approximants is established.

By imposing a further condition on the kernel $k(z, u)$, these approximants become upper and lower bounds for $g(z)$, as is shown in the next theorem. But first we require a lemma.

LEMMA. With the definitions and notation of Equation (2.17),

$$\int_\Gamma w^{m-1} \left\{ [M+J+1/M+1]_f\left(\frac{-1}{w}\right) - [M+J/M]_f\left(\frac{-1}{w}\right) \right\} dw = 0$$

$$\text{for} \quad m = 0, 1, 2, \ldots, 2M+J. \quad (2.19)$$

Proof. We notice first that the Padé accuracy-through-order conditions guarantee that the integral in (2.19) is well defined at $w = \infty$. Hence Γ may be deformed to a contour Γ' enclosing $w = 0$ and the poles of the approximants. Using (2.12), (2.13), and (2.14), we see that, for $m = 0, 1, \ldots, 2M+J$,

$$\frac{1}{2\pi i} \int_{\Gamma'} w^{m-1} [M+J/M]_f\left(\frac{-1}{w}\right) dw$$

$$= \frac{1}{2\pi i} \int_{\Gamma'} w^{m-1} \left\{ \sum_{j=0}^{J} \beta_j w^{-j} + \sum_{i=1}^{M} \frac{\alpha_i}{1 - u_i/w} \right\} dw$$

$$= f_m. \quad (2.20)$$

The other term of (2.19) is treated similarly, and the lemma is proved.

THEOREM 1.2.2. *The inequalities*

$$(-)^{J+1} \left\{ G^{[M+J+1/M+1]}(x) - G^{[M+J/M]}(x) \right\} \geqslant 0, \quad (2.21)$$

$$(-)^{J+1} \left\{ G^{[M+J/M]}(x) - G^{[M+J+1/M-1]}(x) \right\} \geqslant 0, \quad (2.22)$$

$$G^{[M/M]}(x) \geqslant g(x) \geqslant G^{[M-1/M]}(x) \quad (2.23)$$

for all x real and positive, all $M \geqslant 1$ and $J \geqslant -1$, hold if and only if $k(x, u)$ satisfies

$$\left(-\frac{\partial}{\partial u}\right)^j k(x, u) \geqslant 0 \qquad (2.24)$$

for all real nonnegative x and u and all $j = 0, 1, 2, \ldots$.

Proof. First, we show that (2.21) follows from (2.24). Consider a function $K(w)$ which is analytic within a distance Δ from the real w-axis. Let

$$I_K = \frac{1}{2\pi i} \int_\Gamma \frac{K(w)}{w} \left\{ [M+J+1/M+1]_f\left(\frac{-1}{w}\right) - [M+J/M]_f\left(\frac{-1}{w}\right) \right\} dw.$$

$$(2.25)$$

Since $f(z)$ is a Stieltjes series, the zeros of

$$w^{2M+1} Q^{[M+J+1/M+1]}(-1/w) Q^{[M+J/M]}(-1/w)$$

occur at points $w = w_i$, $i = 1, 2, \ldots, 2M+1$, lying on the real w-axis. Using Newton's formula (1.7), we may express $K(w)$ as

$$K(w) = \sum_{i=1}^{2M+1} K[w_1, w_2, \ldots, w_i] \prod_{k=1}^{i-1} (w - w_k)$$

$$+ K[w_1, w_2, \ldots, w_{2M+1}, w] \prod_{k=1}^{2M+1} (w - w_k).$$

Substituting this into (2.25), the lemma shows that

$$I_K = \frac{1}{2\pi i} \int_\Gamma \frac{K[w_1, w_2, \ldots, w_{2M+1}, w]}{w}$$

$$\times \prod_{k=1}^{2M+1} (w - w_k) \left\{ [M+J+1/M+1]_f\left(\frac{-1}{w}\right) - [M+J/M]_f\left(\frac{-1}{w}\right) \right\} dw.$$

Using (I.3.5.18), it follows that

$$I_K = \frac{(-)^{J+1}[C(M+J+1/M+1)]^2}{2\pi i}$$

$$\times \int_\Gamma \frac{w^{-2M-J-2} \prod_{k=1}^{2M+1} (w-w_k) K[w_1, w_2, \ldots, w_{2M+1}, w]}{Q^{[M+J+1/M+1]}\left(\frac{-1}{w}\right) Q^{[M+J/M]}\left(\frac{-1}{w}\right)} \, dw$$

$$= \frac{(-)^{J+1}}{2\pi i} \frac{C(M+J+1/M+1)}{C(M+J/M)}$$

$$\times \int_\Gamma \frac{w^{-J-1} \prod_{k=1}^{2M+1} \left(1-\frac{w_k}{w}\right) K[w_1, w_2, \ldots, w_{2M+1}, w]}{B^{[M+J+1/M+1]}\left(\frac{-1}{w}\right) B^{[M+J/M]}\left(\frac{-1}{w}\right)} \, dw$$

$$= \frac{(-)^{J+1}}{2\pi i} \frac{C(M+J+1/M+1)}{C(M+J/M)} \int_\Gamma w^{-J-1} K[w_1, w_2, \ldots, w_{2M+1}, w] \, dw$$

Since $f(z)$ is a Stieltjes series, $C(M+J+1/M+1)/C(M+J/M)$ is positive if J is odd and negative otherwise (see Exercise 3 of Part I, Section 5.1). Using (1.5), (1.7), and Rolle's theorem, we have

$$\frac{(2M+J+1)!}{2\pi i} \int_\Gamma w^{-J-1} K[w_1, w_2, \ldots, w_{2M+1}, w] \, dw = \left(\frac{d}{du}\right)^{2M+J+1} K(u) \bigg|_{u=\bar{w}},$$

where $\bar{w}$ lies on the real w-axis. Now we invoke (2.17), (2.25) to deduce that

$$(-)^{J+1}\{G^{[M+J+1/M+1]}(x) - G^{[M+J/M]}(x)\} = p\left(\frac{-\partial}{\partial u}\right)^{2M+J+1} k(x, u) \bigg|_{u=\bar{w}},$$

where p is positive. Hence we see that (2.24) is a sufficient condition to prove (2.21).

The other inequalities (2.22), (2.23) are proved similarly, and the convergence of Theorem 1.2.1 justifies the appearance of $g(x)$ in (2.23) as stated. The converse is based on the linearity of (2.5) in $d\phi(u)$, and we refer to Baker [1970] for the details.

Example 1.

$$k(z,u) = \frac{1}{1+zu}.$$ (2.26)

The condition of Theorem (1.2.1) and Equation (2.24) are satisfied. The Baker–Gammel approximants are given by (2.7) as

$$G^{[M+J/M]}(z) = \sum_{j=0}^{J} \beta_j (-z)^j + \sum_{i=1}^{M} \frac{\alpha_i}{1+zu_i}.$$

We see that (2.26) reduces to the special case of $[M+J/M]$ Padé approximants to a Stieltjes series $g(z)$ given by (2.5).

Example 2.

$$k(z,u) = e^{-zu}.$$ (2.27)

The condition of Theorem 1.2.1 and Equation (2.24) are satisfied. The Baker–Gammel approximants are given by (2.7) as

$$G^{[M+J/M]}(z) = \sum_{j=0}^{J} \beta_j \frac{(-z)^j}{j!} + \sum_{i=1}^{M} \alpha_i e^{-z\sigma_i}$$

These approximants are especially useful when the given function is exponentially damped.

Example 3.

$$k(z,u) = \left[1 + \frac{zu}{n}\right]^{-n}$$ (2.28)

The conditions of Theorem 1.2.1 and Equation (2.24) are satisfied for $n > 0$. The Baker–Gammel approximants are

$$G^{[M+J/M]}(z) = \sum_{j=0}^{J} \beta_j \frac{\Gamma(n+j)}{\Gamma(n)} \left(\frac{-z}{n}\right)^j + \sum_{i=1}^{M} \frac{\alpha_i}{(1+zu_i/n)^n}.$$

This class of functions interpolates between Example 1 ($n=1$) and Example 2 ($n=\infty$).

Example 4.

$$k(z,u) = (1 + 2uz + u^2)^{-1/2}.$$ (2.29)

The conditions of Theorem 1.2.1 are satisfied. To establish (2.24) in this case, we let $x>1$ and $u>0$. Then the two roots of $1+2ux+u^2=0$ occur at u_1, u_2 such that $-\infty<u_1<u_2<0$. Cauchy's theorem leads to an integral representation of the kernel as

$$k(z,u)=\frac{1}{\pi}\int_{u_1}^{u_2}\frac{dw}{w-u}\frac{1}{\sqrt{-(1+2wz+w^2)}},$$

and the condition (2.24) is established. Thus (2.29) is a Baker–Gammel kernel. The approximants

$$G^{[M+J/M]}(x)=\sum_{j=0}^{J}\beta_j P_j(x)+\sum_{i=1}^{M}\alpha_i\left(1+2u_ix+u_i^2\right)^{-1/2}$$

give converging upper and lower bounds for $x>1$ to theorem 7.2.2.

Clearly, the preceding analysis of the Padé–Legendre series (2.1) generated by the kernel (2.29) extends to a wider class of orthogonal polynomials, but no general theory has yet been established.

Example 5 (Padé-Borel approximation).   Consider the kernel

$$k(z,u)=\int_0^{\infty}\frac{e^{-t}dt}{1+uzt^p}\qquad\text{for}\quad p>0,$$

which has the asymptotic expansion

$$k(z,u)\simeq\sum_{j=0}^{\infty}(-uz)^j(pj)!,$$

and satisfies the requirements of Theorem 1.2.1 and (2.24). Equation (2.10) takes the form

$$g(z)=\sum_{j=0}^{\infty}c_j(pj)!z^j$$

in this case, and so we need to consider Padé approximants for

$$f(z)=\sum_{j=0}^{\infty}c_jz^j.$$

The difference between these power series of $g(z)$ and $f(z)$ is that a "convergence factor" of $1/(pj)!$ has appeared. The Padé–Borel approximants of $g(z)$ take the form

$$g^{[M+J/M]}(z) = \sum_{j=0}^{\infty} \beta_j (pj)! z^j + \int_0^{\infty} e^{-t} \sum_{i=1}^{M} \frac{\alpha_i}{1+u_i z t^p} dt$$

$$= \int_0^{\infty} e^{-t} f^{[M+J/M]}(z t^p) dt$$

The Padé–Borel method was first used by Graffi et al. [1970] in an application to the anharmonic oscillator; the approach we adopt here is based on that of EPA [p. 287].

A recent extension of the ideas of this section is appropriate when approximants are needed for the series (2.10),

$$g(z) = \sum_{m=0}^{\infty} f_m P_m(z) \tag{2.30}$$

and information is available *a priori* about the asymptotic form of the coefficients f_m [Baker and Gubernatis, 1981]. The scheme of Baker–Gammel approximants can easily be modified to incorporate this extra information. For example, a number of coefficients of the series

$$g(z) = \sum_{m=0}^{\infty} f_m P_m(z) \tag{2.31}$$

might be known, and additionally it might be known that numbers $\{\gamma_n\}$ are given such that

$$f_n \simeq \gamma_n \qquad \text{as} \quad n \to \infty. \tag{2.32}$$

In fact, it turns out that it is only necessary to know the asymptotic form of the ratio

$$R_n = f_{n+1} f_{n-1} / f_n^2, \tag{2.33}$$

as will become clear shortly. Note the similarity of the hypothesis (2.32) to that of Levin's method of Section 1.4.

We present a method for obtaining asymptotic Baker–Gammel approximants, omitting the preconditions for existence and convergence. Equations (2.12), (2.13) are replaced by the hypothesis that, for $J \geqslant -1$, the coefficients $\{f_n\}$ can be expressed as

$$f_m = \gamma_m \left(\beta_m + \sum_{i=1}^{M} \alpha_i u_i^m \right), \qquad m = 0, 1, \ldots, J, \tag{2.34}$$

$$f_m = \gamma_m \left(\sum_{i=1}^{M} \alpha_i u_i^m \right), \qquad m = J+1, J+2, \ldots, 2M+J. \tag{2.35}$$

We then suppose that (2.35) is approximately true for $m = 2M, 2M+1, \ldots$. We also see that multiplying each γ_m by a factor of αh^m for $m = 0, 1, 2, \ldots,$ $2M-1$ leaves the representation (2.35) invariant, showing that the asymptotic form of the ratio R_n in (2.33) is all that need be specified to define the asymptotic approximants.

Equation (2.9) is replaced by

$$k^{(a)}(z, u) = \sum_{m=0}^{\infty} u^m \gamma_m k_m(z). \tag{2.36}$$

It is understood that $k^{(a)}(z, u)$ is defined by analytic continuation in u where necessary. Then we find that the approximants for $g(z)$, defined by the coefficients $\{f_m, m = 0, 1, \ldots, 2M+J\}$ of (2.30) and (2.35)–(2.37) are

$$g^{[M+J/M]}(z) = \sum_{i=1}^{M} \alpha_i k^{(a)}(z, u_i) + \sum_{m=0}^{J} \gamma_m \beta_m k_m(z). \tag{2.37}$$

Exploitation of the information expressed by (2.32) follows the spirit of Levin's method of Section 1.4, and this method also bears resemblance to that of Common and Stacey [1979a].

Example. Suppose that approximants are required for the series

$$g(z) = \sum_{m=0}^{\infty} f_m P_m(z),$$

and it is known that

$$f_n \simeq 6(n+1)3^n \qquad \text{as} \quad n \to \infty.$$

We define $\gamma_n = n+1$ for $n = 0, 1, 2, \ldots,$ and deduce from (2.36) that the kernel appropriate for (2.37) is

$$k^{(a)}(z, u) = (1 - uz)(1 - 2uz + u^2)^{-3/2}.$$

Exercise The Le Roy function is given by

$$L_\zeta(x) = \int_0^\infty \exp\left[-(t + xt^\zeta)\right] dt.$$

Verify that $k(z, u) = L_\zeta(zu)$ satisfies the conditions of Theorem 1.2.1 and Equation (2.24) for $\zeta \geq 0$, and investigate the special cases $\zeta = 0$, $\zeta = 1$.

1.3 Series Analysis

The fundamental motive for studying Padé approximants is the extraction of information about a function from the first few terms of its power series. Before the advent of Padé approximants in theoretical physics, various methods with the collective name of series analysis were used. We start at the beginning with the ratio method, take the Padé method for granted, and proceed to Gammel–Guttmann–Gaunt–Joyce approximants* (abbreviated G³J approximants). Next we come to quadratic approximants and finally to an embryonic method of Levin.

The methods we are about to discuss depend on having either some knowledge of what is a "reasonable functional form" or else of the "general behavior of the power series coefficients". The methods of this section are working techniques for applied scientists. They work in several contexts, and it is instructive to review, very briefly, one context so as to see the kind of intuitive information which proves useful.

In statistical mechanics there are a large number of problems for which the first few terms of the power series may be obtained exactly (they are integers), while the exact solution is unobtainable. The three-dimensional Ising model is a good example. The first few terms of the series of expansion of a thermodynamic quantity, such as the magnetic susceptibility, are generated, and the series is then analysed to determine the behavior of the thermodynamic quantity in question. In statistical mechanics the emphasis is on phase transitions, such as the abrupt onset of ferromagnetic order from the disordered paramagnetic state at the critical temperature. It was expected that the magnetic susceptibility $f(z)$, a function of $z = T^{-1}$, the inverse temperature, would be given by

$$f(z) \approx A(1 - \mu z)^{-\gamma}, \tag{3.1}$$

where $1/\mu$ is the "critical point", γ is the "critical exponent", and $\approx$ means

---

*These approximants are so named because John Gammel and David Gaunt talked at the Canterbury Conference about their work, and discovered that day that A. J. Guttmann and G. S. Joyce were publishing a letter to the same effect. They are also called integral curve approximants.

that the functional form is reasonably good near $\mu z = 1$ and no more. It is known that

$$f(z) = c_0 + c_1 z + c_2 z^2 + c_3 z^3 + \cdots, \tag{3.2}$$

and the problem can be expressed as one of fitting

$$c_n \simeq A \frac{(\gamma+1)\cdots(\gamma+n)}{n!} \mu^n, \qquad n = 0, 1, 2, \ldots, N$$

for the best values of A, μ, and γ. The ratio method [Domb and Sykes, 1961] consists of fitting the ratios

$$r_n \equiv \frac{c_n}{c_{n-1}} \approx \mu \left(1 + \frac{\gamma}{n}\right) \tag{3.3}$$

by suitable means, such a graph of r_n versus n^{-1}. The ratio method is simple but limited in scope. It was replaced by the logarithmic derivative method [Baker, 1961], which consists of noting that (3.1) is equivalent to

$$g(z) \equiv \frac{d}{dz} \ln f(z) \approx \frac{\gamma\mu}{1 - \mu z}.$$

Thus a rational fraction is expected to be the dominant part of $g(z)$, and the values of γ and μ are easily obtained from the pole and residue of the Padé approximants of $g(z)$. This method remains satisfactory because the presumed functional form has been fully exploited but does not constrain the approximation method. Properly used, this method is simple, reliable, and elegant (cf. Part I, Section 2.3).

$G^3 J$ approximants appear as a generalization of the statement that (3.3) is tantamount to the requirement that c_i obey a recurrence relation

$$nc_n - \mu(n+\gamma)c_{n-1} \approx 0$$

which may be rewritten as

$$A_{01} n c_n + \left[A_{11}(n-1) + A_{10}\right] c_{n-1} \approx 0$$

with $A_{01} = 1$, $A_{11} = -\mu$ and $A_{10} = -\mu(1+\gamma)$. This is generalized to become the Mth-order recurrence relation [Guttmann and Joyce, 1972]

$$R_{2M}(c_n) \equiv \sum_{i=0}^{\min(M,n)} \left\{A_{i2}(n-i)^2 + A_{i1}(n-i) + A_{i0}\right\} c_{n-i} = 0 \tag{3.4}$$

which is valid for $n = 1, 2, 3, \ldots$ with the conditions $A_{02} = 1$, $A_{00} = 0$. The unknown coefficients

$$\{A_{01}; A_{i2}, A_{i1}, A_{i,0}, i=1,2,\ldots, M\} \tag{3.5}$$

are determined by the $3M+1$ linear equations

$$R_{2M}(c_n)=0, \qquad n=1,2,\ldots,3M+1. \tag{3.6}$$

Thus (3.6) determines (3.5), and (3.5) determines an entire sequence of coefficients c_i which are given by (3.2) to order $3M+1$ and define an approximating function

$$\psi_M(z)= \sum_{i=0}^{\infty} c_i z^i$$

which agrees with the given expansion to order z^{3M+1}. It may be verified that $\psi_M(z)$ satisfies the ordinary homogeneous differential equation

$$Q(z)\frac{d^2\psi_M}{dz^2} + R(z)\frac{d\psi_M}{dz} + S(z)\psi_M =0, \tag{3.7}$$

where

$$Q(z)=z \sum_{i=0}^{M} A_{i2} z^i,$$

$$R(z)= \sum_{i=0}^{M} (A_{i2} +A_{i1})z^i,$$

and

$$S(z)= \sum_{i=0}^{M-1} A_{i+1,0} z^i.$$

If $Q(z)$ vanishes either once or twice at $z=z_0$ — by which we mean that

$$Q(z)=C_1(z-z_0)+O((z-z_0)^2), \qquad C_1\neq0,$$

or

$$Q(z)=C_2(z-z_0)^2+O((z-z_0)^3), \qquad C_2\neq0$$

— and $S(z_0)\neq0$, then z_0 is called a regular singular point of the differential equation. Depending upon the behavior of $R(z)$ at $z=z_0$, the following functional forms are possible for one of the solutions of (3.7):

$$\psi_M(z) = (z - z_0)^\rho \phi_1(z) + \phi_2(z)$$

(3.8a)

or

$$\psi_M(z) = \ln(z - z_0)\phi_1(z) + \phi_2(z)$$

(3.8b)

or

$$\psi_M(z) = (z - z_0)^\rho \ln(z - z_0)\phi_1(z) + \phi_2(z)$$

(3.8c)

or

$$\psi_M(z) = \ln(z - z_0)\phi_1(z) + (z - z_0)^{-\rho}\phi_2(z) + \phi_3(z),$$

(3.8d)

where $\phi_1(z)$, $\phi_2(z)$, and $\phi_3(z)$ are analytic at $z = z_0$, p is an arbitrary positive integer, and ρ is nonintegral. If $Q_M(z_1) \neq 0$, then z_1 is a regular point of the differential equation and $\psi_M(z)$ is analytic at $z = z_1$. Equations (3.8a–d) show the types of singularity which can be produced accurately by the G^3J scheme defined by (3.4). But the scheme is clearly much more general in its scope, and the inhomogeneous equation

$$Q(z)\frac{d^2\psi_M}{dz^2} + R(z)\frac{d\psi_M}{dz} + S(z)\psi_M = T(z)$$

has also been used in a similar framework. The G^3J approximants satisfy the accuracy-through-order criterion, but are not necessarily rational functions. Their importance is that they allow intuition about the functional form—in other words, its presumed analytic structure near the critical point —to become an integral part of the formulation of the solution.

It is rarely clear in advance which values should be specified for the allowed degrees of the polynomials $Q(z)$, $R(z)$, $S(z)$, and $T(z)$. However, there is one equation of this type for which there is a satisfactory answer [Hunter and Baker, 1979]: the equation

$$R_N(z)\frac{d\psi}{dz} + S_M(z)\psi = T_L(z),$$

where $R_N(z)$, $S_M(z)$, and $T_L(z)$ are polynomials of orders $N = M + 2$, M, and $L = M$ respectively. As degenerate special cases, we note that if $R_N(z) = 0$, the approximant is a diagonal Padé approximant, and if $T_L(z) = 0$, the approximant is a D-log Padé approximant. The important feature of this equation is that it preserves the homographic invariance property. Suppose that we use $3M + 4$ terms of the given series $f(z) = \sum_{i=0}^{\infty} c_i z^i$ to derive a G^3J

approximant which is $\psi_f(z)$. If we make a change of variable $z=w/(a+bz)$, we define

$$g(w)=f(z)=\sum_{i=0}^{\infty} g_i w^i,$$

for which we can obtain a new G³J approximant which is $\psi_g(w)$. The invariance theorem states that the approximant is invariant under the change of variable:

$$\psi_g(w)=\psi_f(z).$$

Ideas similar to the previous ones motivated the introduction of quadratic approximants [Shafer, 1974]. They are a useful growth from Padé's general problem of finding n polynomials $A_1(z), A_2(z),\ldots, A_n(z)$ of degrees $\mu_1, \mu_2,\ldots, \mu_n$ which satisfy

$$A_1(z)f_1(z)+A_2(z)f_2(z)+\cdots+A_n(z)f_n(z)=O(z^{\mu_1+\mu_2+\cdots+\mu_n+n-1}),$$

where $f_i(z)$, $i=1,2,\ldots, n$, are given functions; see the selected bibliography on Hermite-Padé approximation. If $f(z)$ is given and if $Q(z)$, $R(z)$, and $S(z)$ are polynomials of orders q, r, and s which satisfy

$$Q(z)[f(z)]^2+2R(z)f(z)+S(z)=O(z^{q+r+s+2}), \tag{3.9}$$

then the quadratic equation

$$Q(z)[\psi(z)]^2+2R(z)\psi(z)+S(z)=0 \tag{3.10}$$

is easily solved, and a solution is

$$\psi(z)=\frac{-R(z)+\sqrt{[R(z)]^2-S(z)Q(z)}}{Q(z)} \tag{3.11a}$$

$$=\frac{-S(z)}{R(z)+\sqrt{[R(z)]^2-S(z)Q(z)}}. \tag{3.11b}$$

Since $Q(z)$, $R(z)$, and $S(z)$ are polynomials, $\psi(z)$ is analytic except for a finite number of poles and branch points of square-root type. The correct branch of $\psi(z)$ must also be assigned; normally this branch is clear from the context, as in the following

Example.

$$f(z)=\tan^{-1}z.$$

Since $\tan^{-1}z$ is an odd function which vanishes at $z=0$, (3.9) takes the form

$$(1+\alpha z^2)[f(z)]^2+\beta zf(z)+\gamma z^2=O(z^8).\qquad(3.12)$$

The given expansion is

$$f(z)=z-\tfrac{1}{3}z^3+\tfrac{1}{5}z^5-\tfrac{1}{7}z^7+O(z^8);$$

therefore

$$[f(z)]^2=z^2-\tfrac{2}{3}z^4+\tfrac{23}{45}z^6+O(z^8),$$

and substitution in (3.12) gives three simultaneous linear equations for α, β, and γ with the solution

$$\alpha=\tfrac{5}{3},\qquad \beta=3,\quad\text{and}\quad\gamma=-4.$$

From (3.11b), the quadratic approximant of type $[q,r,s]=[2,2,2]$ is

$$\psi(z)=\frac{8z}{3+\left(25+\tfrac{80}{3}z^2\right)^{1/2}}=\tan^{-1}(z)+O(z^8).$$

This approximation has several attractive features. For x real, $\tan^{-1}x$ is well defined and $\tan^{-1}(\infty)=\pi/2=1.57\ldots$ is approximated by $\psi(\infty)=8\sqrt{\tfrac{3}{80}}=1.549\ldots$. For $z=iy$ pure imaginary,

$$\tan^{-1}(iy)=\tfrac{1}{2}\ln\frac{1+iy}{1-iy}=i\tanh^{-1}y,$$

and therefore

$$\tanh^{-1}y=\frac{1}{2i}\ln\frac{1+iy}{1-iy}=\frac{8y}{3+\left(25-\tfrac{80}{3}y^2\right)^{1/2}}+O(y^8),$$

showing the branch points of the quadratic approximants at $y^2=\tfrac{15}{16}$. Thus the global qualities of the quadratic approximants are apparent: we have accurate approximation on the entire real axis and an accurately located singularity on the imaginary axis. Of course, $\tan^{-1}z$ is related to a Stieltjes function, and so the standard Padé method is known to converge systematically in this case; see (I.4.6.5).

There are some series for which the Padé method is an inappropriate choice. Various examples of power series have already been given, such as noisy series, and series based on random numbers. Another example, in this case an ordinary series, is $\Sigma_{i=0}^{\infty} c_i$ where $f(z) = \Sigma_{i=0}^{\infty} c_i z^i$ has a branch point at $z = 1$ and yet $f(1)$ is well defined. One expects the poles of the Padé approximants to accumulate at $z = 1$ and convergence to be slow. For the present, we simplistically regard the Padé method for series summation as the hypothesis that the series is generated by N geometric components

$$c_n = \sum_{i=1}^{N} \alpha_i (r_i)^n, \qquad n = 0, 1, 2, \ldots,$$

where each $|r_i| < 1$ and the partial sums define the sequences

$$S_n = \sum_{i=1}^{N} \alpha_i \frac{1 - r_i^{n+1}}{1 - r_i} = S_\infty - \sum_{i=1}^{N} \left(\frac{\alpha_i r_i}{1 - r_i} \right) r_i^n.$$

In this case, the $[L/M]$ is exact provided $L, M \geq N$. If this hypothesis is inapplicable, one may consider $G^3 J$ approximants for $f(z) = \Sigma_{i=0}^{\infty} c_i z^i$ and evaluate at $z = 1$, especially if the locations of the singularities of $f(z)$ are known. A quite different approach is based on the hypothesis that S_n is a smooth function of the variable $w = 1/n$ on the interval $0 \leq w \leq 1$. The problem is then to interpolate the value of S_n at $n = \infty$. A standard approach is to use rational interpolation (N-point Padé approximation) to interpolate to $w = 0$; any reliable method from Section 1.1 is usually very satisfactory in this context. An interesting approach by Levin [1973] and developed by Sidi [1979, 1980a] is based on the hypothesis that

$$S_n = S_\infty + R(n) \sum_{i=0}^{N} \gamma_i n^{-i}, \qquad n = 1, 2, \ldots, \tag{3.13}$$

where $R(n)$ is a simple function chosen so that $S_\infty + R(n)$ matches the partial sums S_n of the given series as well as possible. The art of the method is to choose the series which constitute S_1 as shrewdly as possible; the choice $R(r) = \int_r^\infty c_n \, dn$ is recommended. The parameters γ_i ($i = 0, 1, 2, \ldots, N$) and S_∞ are determined by $N + 1$ simultaneous linear equations, and so (3.13) determines S directly. In fact, the analysis need require no more than inversion of a Vandermonde matrix (see Exercise 3 of Part I, Section 6.2).

Example. Sum the series

$$\sum_{i=1}^{\infty} c_i = \sum_{i=1}^{\infty} \frac{1}{i^2}. \tag{3.14}$$

The Padé approach suggests that we consider the associated function

$$f(z) = \sum_{i=1}^{\infty} c_i z^i = -\int_0^z \frac{1}{y} \ln(1-y)\, dy, \tag{3.15}$$

which is a dilogarithm. The integrand has a logarithmic branch point at $y = 1$, and so the integral has an end-point singularity of logarithmic type at $z = 1$. Replacing (3.15) by a contour integral, with a little ingenuity one finds that $f(1) = \pi^2/6$, and so we use (3.14) as a test of our numerical methods. The briefest inspection of (3.14) shows its convergence to be slow. Thus we are required to evaluate a power series on its circle of convergence, and so we expect difficulties with a naive application of the Padé method via the ε-algorithm. G^3J approximants would be satisfactory in principle. We will compare four methods: the ε-algorithm and its first iteration, Thiele's rational interpolation method, and Levin's asymptotic expansion method.

Table 1. Estimates of the Sum of the Series (3.14)

n	S_n	Method 1 ε-algorithm ($[l/l]$ Padé)	Method 2 Iterated ε-algorithm	Method 3 Rational interpolation (n-point Padé)	Method 4 Levin's interpolation
1	1.0	1.0	1.0	1.0	1.0
3	1.361 111	1.45	—	1.65	1.625
5	1.463 611	1.551 617	1.581 258	1.644 895	1.644 965
7	1.511 797	1.590 305	—	1.644 934	1.644 935
9	1.539 768	1.609 087	1.629 670	1.644 934	1.644 934

The results in Table 1 show the estimates of S_∞ using $n = 2l+1$ terms of the series (3.14). It is clear that interpolation in $w = n^{-1}$ gives geometric convergence, which is quite satisfactory; for the reasons given, the Padé method is unacceptably slow, and iteration scarcely improves its convergence sufficiently.

We conclude this section by emphasizing that the various methods are special ones (hopefully not too special), and we stress that the methods should only be used in their natural contexts. For further details about G^3J approximants, we refer to Gammel [1973], Joyce and Guttmann [1973], and Guttmann [1975a, b]. For further details about selected methods of series analysis, we refer to Gray, Atchison, and McWilliams [1971], Smith and Ford [1979] and Brezinski, [1980].

Exercise 1. Verify that the entries for method 1 and method 3 in the second row ($n = 3$) of Table 1 are correct.

Exercise 2. Prove that

$$\int_0^1 \frac{1}{y} \ln(1-y)\, dy = \frac{\pi^2}{6}.$$

Hint: First evaluate

$$\int_{-\infty}^{\infty} \frac{\ln|w|}{1-w^2} \, dw$$

using the residue theorem.

1.4 Multivariable Approximants

A natural problem is the generalization of Padé approximants to more than one variable. It turns out that the problems associated with many variables have the same kind of solution as the two-variable problems, and we confine our attention to this case for ease of exposition. First we consider various schemes for the formation of two-variable rational approximants, and then we consider more general classes of approximating functions. Let us assume that we are given the coefficients of the series expansion

$$f(x, y) = \sum_{i=0}^{\infty} \sum_{j=0}^{\infty} c_{ij} x^i y^j. \tag{4.1}$$

The problem consists of defining lattice spaces $\mathfrak{N}$ and $\mathfrak{D}$ and polynomials

$$A(x, y) = \sum_{i,j \in \mathfrak{N}} a_{ij} x^i y^j \tag{4.2}$$

and

$$B(x, y) = \sum_{i,j \in \mathfrak{D}} b_{ij} x^i y^j, \tag{4.3}$$

so that

$$f(x, y) = \frac{A(x, y)}{B(x, y)} + \sum_{i=0}^{\infty} \sum_{j=0}^{\infty} e_{ij} x^i y^j, \tag{4.4}$$

where as many coefficients e_{ij} as possible are zero. We have taken the numerator and denominator coefficients to lie in lattice spaces $\mathfrak{N}$ and $\mathfrak{D}$, and we require that $e_{ij} = 0$ for $i, j \in \mathfrak{E}$, the equality lattice space.

Taking $b_{00} = 1$ as part of the definition, this scheme is normally determinate if

$$\dim(\mathfrak{E}) = \dim(\mathfrak{N}) + \dim(\mathfrak{D}) - 1.$$

This analysis provides the foundation for a variety of approximation schemes. These schemes are only useful if their properties are known, and the most systematic developments are known as the Canterbury approximants (or generalized Chisholm approximants). These approximants have many properties; they satisfy accuracy-through-order conditions and reduce to Padé approximants if either x or y is zero. The original and simplest of the Canterbury approximants is the Chisholm approximant [Chisholm, 1973], defined by writing (4.2) and (4.3) as

$$A^{[L/L]}(x, y) = \sum_{i=0}^{L} \sum_{j=0}^{L} a_{ij} x^i y^j \tag{4.5}$$

$$B^{[L/L]}(x, y) = \sum_{i=0}^{L} \sum_{j=0}^{L} b_{ij} x^i y^j, \qquad b_{00} = 1. \tag{4.6}$$

The lattice spaces $\mathfrak{N}$ and $\mathfrak{D}$ corresponding to (4.5) and (4.6) are shown in Figure 1. Then one has equality at order $x^\alpha y^\beta$ if

$$\sum_{i=0}^{\alpha} \sum_{j=0}^{\beta} b_{ij} c_{\alpha-i, \beta-j} = a_{\alpha\beta} \qquad \text{for } (\alpha, \beta) \in \mathfrak{N} \tag{4.7}$$

and

$$\sum_{i=0}^{\min(\alpha, L)} \sum_{j=0}^{\min(\beta, L)} b_{ij} c_{\alpha-i, \beta-j} = 0 \qquad \text{for } (\alpha, \beta) \in \mathfrak{E}, \ (\alpha, \beta) \notin \mathfrak{N}. \tag{4.8}$$

The numerator coefficients are determined by (4.7) once the denominator coefficients are determined. The b_{ij} are determined by (4.8), and the lattice

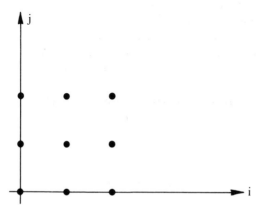

Figure 1. The lattice space $\mathfrak{D}$ required for the [2/2] Chisholm approximant. In this case, the lattice spaces $\mathfrak{N}$ and $\mathfrak{D}$ are identical.

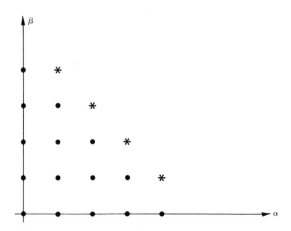

Figure 2. The lattice space ℰ for a [2/2] Chisholm approximant. A star denotes a lattice point corresponding to an equation to be symmetrized.

space ℰ is most simply described by an example. In Figure 2, we show the space for the [2/2] Chisholm approximant.

To determine b_{10} and b_{20}, and obtain the [2/2] Padé approximant when $y=0$, accuracy to order x^3 and x^4 is required, and no higher. In fact, six of the eight equations for the b_{ij} are obtained by requiring accuracy through order x^3, x^4, x^3y, y^3, y^4, and y^3x, as indicated by * in Figure 2. Two further equations are needed. One is obtained by writing down the equations (4.8) for orders x^4y and xy^4 and adding them, and the other by doing the same for orders x^3y^2 and x^2y^3. These two equations are called symmetrized equations. Thus Chisholm approximants always satisfy accuracy-through-order conditions for orders up to $x^{2L-\alpha}y^\alpha$, $\alpha=0,1,\dots,2L$. For symmetric functions, the symmetrizing process becomes a formality, and the approximants are accurate through orders x^{2L}, y^{2L}, and $x^{2L+1-\alpha}y^\alpha$, $\alpha=1,2,\dots,2L$. Provided that the necessary approximants exist, the scheme has the following properties [Chisholm, 1973; Common and Graves-Morris, 1974]:

Chisholm approximants reduce to diagonal Padé approximants if either variable is zero.

They satisfy restricted homographic invariance: let

$$x=\frac{Au}{1+Bu}, \quad y=\frac{Av}{1+Cv}, \quad A\neq 0.$$

If

$$f(x,y)=g(u,v),$$

then

$$[L/L]_f(x,y)=[L/L]_g(u,v).$$

They satisfy duality: If

$$g(x, y) = \frac{1}{f(x, y)},$$

then

$$[L/L]_g(x, y) = \frac{1}{[L/L]_f(x, y)} \qquad [f(0,0) \neq 0].$$

They preserve unitarity:
if $f(x, y)f^*(x, y) = 1$, then $[L/L]_f(x, y) \times \{[L/L]_f(x, y)\}^* = 1$.

They satisfy the factorization rule: if $f(x, y)$ is a product function, the approximant factorizes to a product of Padé approximants. Expressed in formulas, this rule states that if $f(x, y) = g(x)h(y)$, then $[L/L]_f(x, y) = [L/L]_g(x)[L/L]_h(y)$.

Formation of Chisholm approximants commutes with bilinear transformations of the function: let

$$g(x, y) = \frac{A + Bf(x, y)}{C + Df(x, y)}, \qquad C + Df(0,0) \neq 0.$$

Then

$$[L/L]_g(x, y) = \frac{A + B[L/L]_f(x, y)}{C + D[L/L]_f(x, y)}.$$

The proofs of all these properties are based on the accuracy-through-order principle. In particular, the homographic invariance and factorization properties seem to be essential ingredients of a useful scheme.

The general system of Hughes Jones approximants [Hughes Jones, 1976] is defined by

$$A^{[L/M]}(x, y) = \sum_{i=0}^{L_1} \sum_{j=0}^{L_2} a_{ij} x^i y^j$$

$$B^{[L/M]}(x, y) = \sum_{i=0}^{M_1} \sum_{j=0}^{M_2} b_{ij} x^i y^j, \qquad b_{00} = 1.$$

The equality lattice space $\mathcal{E}$ is shown in Figures 3, 4. If $\min(M_1, M_2) \leqslant \min(L_1, L_2)$, the simple situation of Figure 3 applies; otherwise, we have the more complicated situation of Figure 4. The logic of these figures can be understood in terms of the prong method [Hughes Jones and Makinson, 1974]. The coefficients b_{ij} are calculated sequentially in prongs; prongs are

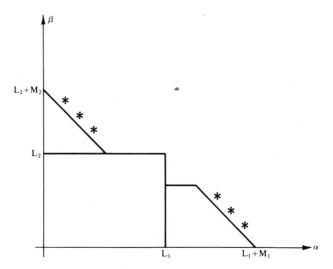

Figure 3. Schematic drawing of the lattice space $\mathfrak{S}$ for a Canterbury approximant with $M_2 < M_1 < \min(L_1, L_2)$ and three symmetrized equations.

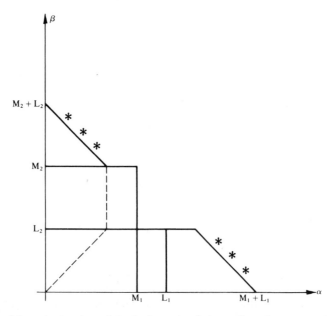

Figure 4. Schematic drawing of the lattice space $\mathfrak{S}$ for a Canterbury approximant with $M_1 < L_1$ and $L_2 < \min(M_1, M_2)$ and three symmetrized equations.

defined to be vectors

$$\mathbf{b}^{(0)} = \left(b_{10}, b_{20}, \ldots, b_{m_1 0}; b_{01}, b_{02}, \ldots, b_{0 m_2} \right),$$

$$\mathbf{b}^{(1)} = \left(b_{21}, b_{31}, \ldots, b_{m_1 1}; b_{12}, b_{13}, \ldots, b_{1 m_2}; b_{11} \right),$$

etc. Calculating $\mathbf{b}^{(0)}$ is equivalent to calculating Padé approximants for $f(x,0)$ and $f(0, y)$, and it turns out that the evaluation of $\mathbf{b}^{(j)}$ only requires values of $\mathbf{b}^{(i)}$, $i = 0, 1, \ldots, j-1$. In summary, the prong method reduces the calculation of b_{ij} to linear algebra with a block lower triangular coefficient matrix.

To define the Canterbury approximants completely, there is the possibility of weighting the symmetrizing equations before adding them. Naturally, for symmetric functions with $c_{ij} = c_{ji}$, the weights are equal. There are two different weighting schemes which treat the variables symmetrically, and each with its own advantages. One weighting scheme [Chisholm and Hughes Jones, 1975] gives full homographic invariance under the changes of variable

$$x = \frac{Au}{1 + Bu}, \qquad y = \frac{Cu}{1 + Du}. \tag{4.9}$$

The other scheme [Graves-Morris and Roberts, 1975] guards against accidental degeneracy by maximizing the numerical stability of the system, and has homographic invariance provided $|A| = |C|$ in (4.9). Fortunately, both schemes lead to similar results in trials.

As stated previously, the preceding ideas may be generalized systematically by geometrical methods and by the prong method to multivariable approximants at the price of algebraic complexity only [Chisholm and McEwan, 1974; Hughes Jones, 1976].

There are two other systems of N-variable rational approximants with distinctive merits. Both lead to higher-order polynomials in numerator and denominator than the accuracy-through-order criterion requires.

First, one may partition the series for $f(x, y)$ by defining [Hillion, 1977a; Watson, 1974]

$$F(x, y; \lambda) = \sum_{i=0}^{\infty} \lambda^i \sum_{k=0}^{i} x^k y^{i-k} c_{k, i-k}.$$

Then $f(x, y) = F(x, y; 1)$, and $[L/M]$ Padé approximants in λ to $F(x, y; \lambda)$ define rational approximants to $f(x, y)$ at $\lambda = 1$. These are very convenient approximants in the presence of many variables, and reduce to $[L/M]$ Padé approximants to $f(x, 0)$ and $f(0, y)$ on the axes. Furthermore, their values may be obtained by the ε-algorithm. They have the disadvantage of requiring a relatively specific set of coefficients for their formation.

Secondly, one may treat $f(x, y)$ as if it defines a moment problem [Alabiso and Butera, 1975]. Let us suppose

$$f(x, y) = \int \frac{\rho(u, t)\, du\, dt}{1 + ux + ty} = \sum_{m, n} \binom{m+n}{m} f_{m, n} x^m y^n, \qquad (4.10)$$

and that $|g\rangle \in \mathcal{H}$, where $\mathcal{H}$ is a Hilbert space with two commuting operators A, B and such that

$$f_{m, n} = \langle g | A^m B^n | g \rangle. \qquad (4.11)$$

These equations (4.10), (4.11) are formally equivalent to the condition that

$$f(x, y) = \langle g | \frac{1}{1 + Ax + By} | g \rangle. \qquad (4.12)$$

The vector

$$|\psi^{(N)}\rangle = \sum_{p=0}^{N} \sum_{q=0}^{p} \psi_{pq}^{(N)} A^{p-q} B^q | g \rangle \qquad (4.13)$$

is an approximation to

$$|\psi\rangle = \frac{1}{1 + Ax + By} | g \rangle \qquad (4.14)$$

provided

$$\langle h_{rs} | 1 + Ax + By | \psi^{(N)} \rangle = \langle h_{rs} | g \rangle \qquad (4.15)$$

for

$$\langle h_{rs} | = \langle g | A^{r-s} B^s$$

with

$$r = 0, 1, \ldots, N \quad \text{and} \quad s = 0, 1, \ldots, r.$$

Equation (4.15) provides $\frac{1}{2} N(N+1)$ linear equations for $\psi_{pq}^{(N)}$, namely

$$\sum_{p=0}^{N} \sum_{q=0}^{p} \psi_{pq}^{(N)} \left(f_{r+p, s+q} + x f_{r+p+1, s+q} + y f_{r+p, s+q+1} \right) = f_{r, s}.$$

This equation determines $|\psi^{(N)}\rangle$ from (4.13) and a rational approximant given by $\langle g | \psi^{(N)} \rangle$ following (4.12). The general result is clear from the example of $N=1$, for which

$$\langle f | \psi^{(1)} \rangle$$

$$= (f_{00} f_{10} f_0) \begin{vmatrix} f_{00} - xf_{10} - yf_{01} & f_{10} - xf_{20} - yf_{11} & f_{01} - xf_{11} - yf_{02} \\ f_{10} - xf_{20} - yf_{11} & f_{20} - xf_{30} - yf_{21} & f_{11} - xf_{21} - yf_{12} \\ f_{01} - xf_{11} - yf_{02} & f_{11} - xf_{21} - yf_{12} & f_{02} - xf_{12} - yf_{03} \end{vmatrix}^{-1} \begin{pmatrix} f_{00} \\ f_{10} \\ f_{01} \end{pmatrix}.$$

This system of approximants converges for strict Stieltjes functions in two variables, but the accuracy-through-order principle and the property of reduction to Padé approximants on the axes are forfeited.

Probably the best of the multivariable approximants described in this section is the one which fits the context of the original problem most closely. For this reason we now consider alternative approximants, not necessarily rational, which may be formed from the coefficients c_{ij} and in some sense approximate

$$f(x, y) = \sum_{i=0}^{\infty} \sum_{j=0}^{\infty} c_{ij} x^i y^j. \tag{4.1}$$

In the context of critical phenomena, we seek a generalization of G^3J approximants (Section 1.3) and D-log Padé approximants (I.2.2.7) for functions having branch points as well as poles. As an example, f might be the specific heat of a substance, depending on temperature and pressure. In

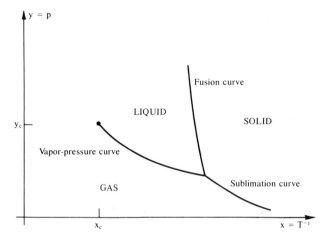

Figure 5. Singularity structure of the specific heat of a fluid in the (x, y) plane as a function of $x = T^{-1}$ and $y = p$. The critical point (x_c, y_c) and the vapor-pressure line representing the liquid–gas transition are shown.

this case we set $x = T^{-1}$ and $y = p$ in (4.1), and investigate the properties of f near the critical point. It is conjectured [Pfeuty et al., 1974] that such functions might behave as

$$f(x, y) \approx (x_c - x)^{-\gamma} Z\left(\frac{y_c - y}{(x_c - x)^{\phi}}\right) \tag{4.16}$$

near the critical point, where γ and ϕ are exponents, and $Z(z)$ is an unknown scaling function. Fisher approximants [Fisher, 1977; Fisher and Kerr, 1977] are designed for problems of the kind just outlined. From the power-series coefficients given, one considers the truncated Maclaurin polynomial

$$f_T(x, y) = \sum_{i, j \in \mathcal{S}} c_{ij} x^i y^j.$$

Its derivatives $\partial f_T(x, y)/\partial x$ and $\partial f_T(x, y)/\partial y$ follow immediately. In the absence of degeneracy, one may construct polynomials $P_{\mathcal{L}}(x, y)$, $Q_{\mathfrak{M}}(x, y)$, and $R_{\mathfrak{N}}(x, y)$ which satisfy

$$P_{\mathcal{L}}(x, y) f_T(x, y) = Q_{\mathfrak{M}}(x, y) \frac{\partial f_T(x, y)}{\partial x} + R_{\mathfrak{N}}(x, y) \frac{\partial f_T(x, y)}{\partial y}$$

$$+ \text{high-order terms.} \tag{4.17}$$

The lattice space $\mathcal{S}$ contains the data coefficients, the lattice spaces $\mathcal{L}$, $\mathfrak{M}$, $\mathfrak{N}$ define the orders of the polynomials $P_{\mathcal{L}}(x, y)$, $Q_{\mathfrak{M}}(x, y)$, and $R_{\mathfrak{N}}(x, y)$, and the scheme is normally determinate if

$$\dim(\mathcal{S}) = \dim(\mathcal{L}) + \dim(\mathfrak{M}) + \dim(\mathfrak{N}) - 1. \tag{4.18}$$

This follows because (4.17) defines a homogeneous linear system of equations for the coefficients of $P_{\mathcal{L}}(x, y)$, $Q_{\mathfrak{M}}(x, y)$, and $R_{\mathfrak{N}}(x, y)$. The precise nature of the lattice spaces $\mathcal{L}$, $\mathfrak{M}$, and $\mathfrak{N}$ may be chosen to suit the problem at hand. Using these polynomials $P_{\mathcal{L}}(x, y)$, $Q_{\mathfrak{M}}(x, y)$, and $R_{\mathfrak{N}}(x, y)$, the Fisher approximant of $f(x, y)$ is defined to be a solution of the partial differential equation

$$P_{\mathcal{L}}(x, y) F(x, y) = Q_{\mathfrak{M}}(x, y) \frac{\partial F(x, y)}{\partial x} + R_{\mathfrak{N}}(x, y) \frac{\partial F(x, y)}{\partial y}. \tag{4.19}$$

In order that $F(x, y)$ may be uniquely defined, a boundary condition must be specified, e.g. the function $F(0, y)$ must be known. $F(0, y)$ might be estimated by a G^3J approximant if its exact value is not available.

Example. The given series coefficients are derived from

$$F(x, y) = \sum_{i=0}^{\infty} \sum_{j=0}^{\infty} c_{ij} x^i y^j = (1 - 2x + y)^{-\gamma} + \theta(1 - x + 2y)^{-2\gamma}, \quad (4.20)$$

where θ, γ are constants. With the choice of lattices $\mathcal{L}$, $\mathcal{M}$, and $\mathcal{N}$ implied by

$$P_{\mathcal{L}}(x, y) = 6\gamma, \qquad Q_{\mathcal{M}}(x, y) = 5 - 7x + 2y, \qquad R_{\mathcal{N}}(x, y) = 4 - 2x - 2y,$$
$$(4.21)$$

$f(x, y)$ is an exact solution of (4.19). With the boundary condition

$$F(0, y) = (1 + y)^{-\gamma} + \theta(1 + 2y)^{-\gamma},$$

(4.20) is the unique solution of (4.19) and (4.21).

Any solution of (4.19) is normally necessarily singular at the point where $Q_{\mathcal{M}}(x, y) = R_{\mathcal{N}}(x, y) = 0$, and this would represent the critical point. For example, any solution of (4.19), (4.21) is normally singular at $x = 1$, $y = 1$. Solutions of (4.19) with the properties shown in Figure 5 require a more sophisticated demonstration.

Fisher approximants are linear in the sense that if F_1, F_2 are solutions of (4.19) for specified $P_{\mathcal{L}}(x, y)$, $Q_{\mathcal{M}}(x, y)$ and $R_{\mathcal{N}}(x, y)$, then $F_1 + F_2$ is also a solution. A class of nonlinear two-variable approximants are the branching approximants which arise as generalizations of Shafer's quadratic approximants. Given the power series (4.1), or at least a sufficient subset $\mathcal{S}$ of its coefficients, the coefficients of the power series of $[f(x, y)]^2$ may be constructed. With preassigned lattice spaces $\mathcal{L}$, $\mathcal{M}$, and $\mathcal{N}$, one may normally construct polynomials $Q_{\mathcal{L}}(x, y)$, $R_{\mathcal{M}}(x, y)$, and $S_{\mathcal{N}}(x, y)$ such that

$$Q_{\mathcal{L}}(x, y)[f(x, y)]^2 + 2R_{\mathcal{M}}(x, y)f(x, y) + S_{\mathcal{N}}(x, y) = 0 \qquad \text{to high order}$$

provided (4.18) is satisfied. As in (3.11a), the quadratic approximant is defined [Chisholm, 1977b] by

$$\psi(x, y) = \frac{-R_{\mathcal{M}}(x, y) + \sqrt{[R_{\mathcal{M}}(x, y)]^2 - S_{\mathcal{N}}(x, y)Q_{\mathcal{L}}(x, y)}}{Q_{\mathcal{L}}(x, y)}. \quad (4.22)$$

The correct branch of $\psi(x, y)$ to represent $f(x, y)$ must also be assigned; this branch is normally clear from the context. The quadratic approximant scheme may be generalized both to higher-order algebraic equations and to different functional equations. We reemphasize that the best N-variable approximant to use is the one which fits the context of the original problem most closely.

For further details about Canterbury approximants, we refer to Lutterodt [1974], to the review by Chisholm [1977a], and to Graves-Morris, Hughes Jones, and Makinson [1974], Chisholm and Graves-Morris [1975], Roberts et al., [1975], Chisholm and Roberts, [1976], Graves-Morris [1977], and Roberts [1977].

For further details of other kinds of multivariable approximants, we refer to Levin [1976], Fisher [1977], and Baker and Moussa [1978]; to Genz [1977]; to Karlsson and Wallin [1977]; to Barnsley and Robinson [1978]; and to Chisholm [1977b, 1978a, b], Short [1978], and Brezinski [1978b].

1.5 Matrix Padé Approximants

Most of this book is about Padé approximants to $\Sigma_{i=0}^{\infty} c_i z^i$, where c_i are real or complex numbers. In fact, most of the properties of Padé approximants discussed have a logical interpretation when c_i are square matrices, and $f(z) = \Sigma_{i=0}^{\infty} c_i z^i$ is also a matrix. We now consider the case where c_i are constant square matrices with real or complex elements. We concentrate on what appear to be the most useful aspects of the generalization. The principal obstacle to be overcome when c_i are matrices is that the order of multiplication is significant. In general, we cannot assume that multiplication of c_i and c_j is commutative. But before analyzing any of the difficulties of defining and constructing matrix Padé approximants, we consider two examples.

Example 1.

$$f(z) = \begin{pmatrix} \cos z & -\sin z \\ \sin z & \cos z \end{pmatrix} = \begin{pmatrix} 1 & 0 \\ 0 & 1 \end{pmatrix} + \begin{pmatrix} 0 & -1 \\ 1 & 0 \end{pmatrix} z + \begin{pmatrix} \frac{1}{2} & 0 \\ 0 & \frac{1}{2} \end{pmatrix} z^2 + \cdots$$

$$= c_0 + c_1 z + c_2 z^2 + \cdots$$

$$= (a_0 + a_1 z)(I + b_1 z)^{-1} + O(z^3). \tag{5.1}$$

This leads to the usual accuracy-through-order conditions:

$$c_0 = a_0,$$

$$c_1 + c_0 b_1 = a_1, \tag{5.2}$$

$$c_2 + c_1 b_1 = 0,$$

and a simple calculation with these matrices shows that

$$[1/1]_f = \begin{pmatrix} 1 & -\tfrac{1}{2}z \\ \tfrac{1}{2}z & 1 \end{pmatrix} \begin{pmatrix} 1 & \tfrac{1}{2}z \\ -\tfrac{1}{2}z & 1 \end{pmatrix}^{-1}.$$

Example 2.

$$f(z) = \begin{pmatrix} 1+z+z^2 & 0 \\ -z & 1+z-z^2 \end{pmatrix} = \begin{pmatrix} 1 & 0 \\ 0 & 1 \end{pmatrix} + \begin{pmatrix} 1 & 0 \\ -1 & 1 \end{pmatrix}z + \begin{pmatrix} 1 & 0 \\ 0 & -1 \end{pmatrix}z^2$$

$$= c_0 + c_1 z + c_2 z^2.$$

Following (5.1), (5.2), we find

$$^R[1/1]_f = \begin{pmatrix} 1 & 0 \\ -2z & 1+2z \end{pmatrix} \begin{pmatrix} 1-z & 0 \\ -z & 1+z \end{pmatrix}^{-1}. \tag{5.3}$$

If, instead of (5.1), we choose to write

$$c_0 + c_1 z + c_2 z^2 = (I + b_1' z)^{-1}(a_0' + a_1' z) + O(z^3),$$

then we need to solve

$$c_0 = a_0',$$
$$c_1 + b_1' c_0 = a_1', \tag{5.4}$$
$$c_2 + b_1' c_1 = 0,$$

with the result that

$$^L[1/1]_f = \begin{pmatrix} 1-z & 0 \\ z & 1+z \end{pmatrix}^{-1} \begin{pmatrix} 1 & 0 \\ 0 & 1+2z \end{pmatrix}. \tag{5.5}$$

The terms of (5.5) are clearly different from those of (5.3); in (5.3), the denominator is on the right, the Padé approximant is right-handed and is denoted by $^R[1/1]$. Correspondingly $^L[1/1]$ is the left-handed Padé approximant. We distinguish the right-handed denominator $^R B^{[1/1]}(z)$ from the left-handed denominator $^L B^{[1/1]}(z)$, and note that in Example 2,

$$^R B^{[1/1]}(z) \neq {}^L B^{[1/1]}(z).$$

DEFINITION. For a series $f(z) = \sum_{i=0}^\infty c_i z^i$, where c_i are $d \times d$ square matrices, we define right-handed Padé approximants by

$$f(z) \, {}^R B^{[L/M]}(z) - {}^R A^{[L/M]}(z) = O(z^{L+M+1}) \tag{5.6}$$

together with $^R B^{[L/M]}(0)=I$, so that $\qquad$ (5.7)

$$^R[L/M]= {}^R A^{[L/M]}(z)\{ {}^R B^{[L/M]}(z)\}^{-1}. \qquad (5.8)$$

Left-handed Padé approximants are defined correspondingly using left inverses.

This definition, incorporating (5.7), is a Baker definition guaranteeing accuracy through order $L+M$. With alternative definitions, denominators with properties such as

$$Q^{[L/M]}(0)=\begin{pmatrix} 1 & 0 \\ 0 & 0 \end{pmatrix}$$

are possible, and are fraught with danger. However, these do not concern us directly. With the Baker-type definition, we have the existence theorem:

THEOREM 1.5.1. *If c_i are $d\times d$ matrices, and provided*

$$\begin{vmatrix} c_{L-M+1} & \cdots & c_L \\ \vdots & & \vdots \\ c_L & \cdots & c_{L+M-1} \end{vmatrix} \neq 0 \qquad (5.9)$$

(which is the determinant of an $Md\times Md$ matrix), then both left- and right-handed Padé approximants exist.

Proof. Following (5.6), $^R B^{[L/M]}(z)$ is defined by

$$\begin{pmatrix} c_L & c_{L-1} & \cdots & c_{L-M+1} \\ c_{L+1} & c_L & \cdots & c_{L-M+2} \\ \vdots & \vdots & & \vdots \\ c_{L+M-1} & c_{L+M-2} & \cdots & c_L \end{pmatrix}\begin{pmatrix} b_1 \\ b_2 \\ \vdots \\ b_M \end{pmatrix}=\begin{pmatrix} -c_{L+1} \\ -c_{L+2} \\ \vdots \\ -c_{L+M} \end{pmatrix}. \qquad (5.10)$$

This equation describes a block structure. The right-hand side has d columns, and so (5.10) represents d sets of simultaneous linear equations with the same $Md\times Md$ coefficient matrix. Thus (5.9) is the condition for a unique solution.

THEOREM 1.5.2. *Left-handed and right-handed Padé approximants are identical but may have different representations.*

Proof. By virtue of the defining equations (5.6), (5.7), and (5.8),

$$^L[L/M]= {}^R[L/M]+O(z^{L+M+1}).$$

Multiply on the right by $^RQ^{[L/M]}(z)$ and on the left by $^LQ^{[L/M]}(z)$, giving

$$^LQ^{[L/M]}(z)\,^RP^{[L/M]}(z) - {}^LP^{[L/M]}(z)\,^RQ^{[L/M]}(z) = O(z^{L+M+1})$$
$$= 0$$

because there are no terms of order z^{L+M+1} on the left-hand side. The theorem follows by division.

This somewhat paradoxical theorem is easily appreciated by noticing that $\frac{1}{2} \times 6$ and $9 \times \frac{1}{3}$ are different representations of the same number. In the matrix context, we return to (5.3).

Example 2 (continued). We construct explicitly the exact inverses of $^RQ(z)$ and $^LQ(z)$ in (5.3) and (5.5):

$$^R[1/1] = \begin{pmatrix} 1 & 0 \\ -2z & 1+2z \end{pmatrix} \begin{vmatrix} \dfrac{1+z}{1-z^2} & 0 \\ \dfrac{z}{1-z^2} & \dfrac{1-z}{1-z^2} \end{vmatrix} = \begin{pmatrix} \dfrac{1}{1-z} & 0 \\ \dfrac{-z}{1-z^2} & \dfrac{1+2z}{1+z} \end{pmatrix},$$

$$^L[1/1] = \begin{vmatrix} \dfrac{1+z}{1-z^2} & 0 \\ \dfrac{-z}{1-z^2} & \dfrac{1-z}{1-z^2} \end{vmatrix} \begin{pmatrix} 1 & 0 \\ 0 & 1+2z \end{pmatrix} = \begin{pmatrix} \dfrac{1}{1-z} & 0 \\ \dfrac{-z}{1-z^2} & \dfrac{1+2z}{1+z} \end{pmatrix},$$

showing that in this case the left- and right-hand Padé approximants are identical, and confirming that each equals the given matrix through order z^2.

THEOREM 1.5.3. *Consider a transformation of the series coefficients defined by*

$$c_i \rightarrow c_i' = vc_iv^{-1},$$

where v is any constant $d \times d$ invertible matrix. This transformation defines an automorphism of the algebra of coefficient matrices, and specifically it defines

$$g(z) = \sum_{i=0}^{\infty} c_i'z^i = v \cdot f(z) \cdot v^{-1}.$$

Then $[L/M]_g(z) = v \cdot [L/M]_f(z) \cdot v^{-1}$, provided either approximant exists.

THEOREM 1.5.4. *If $f(0) = c_0$ is invertible and*

$$g(z) = 1/f(z),$$

then

$$[L/M]_g(z)=\{[M/L]_f(z)\}^{-1},$$

provided either approximant exists.

THEOREM 1.5.5. *Let* L_1, L_2, L_3, L_4 *be constant* $d \times d$ *matrices, and let*

$$g(z)=(L_1 f(z)+L_2)(L_3 f(z)+L_4)^{-1}.$$

Then

$$[M/M]_g=\{L_1[M/M]_f+L_2\}\{L_3[M/M]_f+L_4\}^{-1}$$

provided $L_3 f(0)+L_4$ *is invertible and* $[M/M]_f$ *exists.*

THEOREM 1.5.6 [Gammel and McDonald, 1966]. *Let* $S(z)$ *be unitary, viz.*

$$S(z)\{S(z)\}^{\dagger}=I, \tag{5.11}$$

where † *denotes the Hermitian conjugate, i.e. transposition followed by complex conjugation. Provided the* $[M/M]$ *matrix Padé approximants exist, then they are unitary:*

$$[M/M]_S(z)\{[M/M]_S(z)\}^{\dagger}=I. \tag{5.12}$$

THEOREM 1.5.7. *Let* $w=\alpha z/(\beta z+\gamma)$ *and* $g(w)=f(z)$. *Then* $[N/N]_g(w)=[N/N]_f(z)$ *provided* $\alpha, \gamma \neq 0$ *and either approximant exists.*

Proofs. The proofs are all based in the accuracy-through-order method and are formally identical to those of the scalar case.

An interesting view of matrix Padé approximants derives from a comparison with the consequences of a variational approach. Define a right-handed functional by

$$\mathfrak{R}=\mathfrak{R}(\mu_0,\mu_1,\dots\mu_N)=\sum_{k=0}^{N} c_k \mu_k z^k,$$

a left-handed functional by

$$\mathcal{L}=\mathcal{L}(\lambda_0,\lambda_1,\dots,\lambda_N)=\sum_{j=0}^{N} \lambda_j c_j z^j,$$

and a bilateral Hankel functional by

$$\mathcal{K} = \mathcal{K}(\lambda_0, \lambda_1, \ldots, \lambda_N; \mu_0, \mu_1, \ldots, \mu_N)$$
$$= \sum_{j=0}^{N} \sum_{k=0}^{N} \lambda_j z^j (c_{j+k} - z c_{j+k+1}) z^k \mu_k.$$

The parameters λ_j, μ_k are arbitrary $d \times d$ matrices.

THEOREM 1.5.8. *The variational functions are defined by*

$$[F_1(\lambda_0, \ldots, \lambda_N; \mu_0, \ldots, \mu_N)] = \mathcal{R} \mathcal{K}^{-1} \mathcal{L},$$
$$[F_2(\lambda_0, \ldots, \lambda_N; \mu_0, \ldots, \mu_N)] = \mathcal{R} - \mathcal{K} + \mathcal{L}.$$

These two functionals are identical at their respective stationary points and are then the $[M-1/M]$ matrix Padé approximants of $\sum_{i=0}^{\infty} c_i z^i$.

Space permits neither the proof of this theorem nor a full incursion into the elegant formalism of variational methods with matrix coefficients. In Section 2.8, we derive matrix Padé approximants from the ordinary Schwinger principle in the context of potential theory. In fact, matrix Padé approximants have principally been used in nuclear and elementary-particle physics.

The convergence theory of matrix Padé approximants is somewhat incomplete, and especially so if the matrices are infinite-dimensional matrices. However, if $f(z)$ is an R-operator corresponding to Stieltjes series with infinite matrices as the coefficients c_i, convergence theorems exist.

We turn next to the thorny problems occurring when the c_i are not square matrices. We consider the case where c_i are vectors, or $d \times 1$ matrices. The first problem is to define c_i^{-1}, for which we use Samelson's formula

$$c_i^{-1} = \frac{c_i^*}{\|c_i\|^2} \tag{5.13}$$

and the $\mathcal{L}_2$ norm is natural in this context; furthermore, empirically the $\mathcal{L}_2$ norm works best. Instead of attempting to interpret (5.10), we note that Wynn's algorithm supplemented by (5.13) is all that is required, provided the intermediate approximants exist. The only justification of this procedure can be that it produces reasonable results in idealized circumstances. We consider the situation where the ε-algorithm is used to accelerate the convergence of a sequence of vectors satisfying the recurrence relation

$$\sum_{j=0}^{M} \alpha_j \mathbf{v}_{i+j} = \left(\sum_{j=0}^{M} \alpha_j \right) \mathbf{v}, \qquad i = 0, 1, 2, \ldots, \tag{5.14}$$

where $\alpha_0, \alpha_1, \ldots, \alpha_M$ are constants such that $\Sigma_{j=0}^M \alpha_j \neq 0$. Then the ε-algorithm accelerates the convergence of the partial sums to the expected value

$$\varepsilon_{2M}^{(i)} = \mathbf{v}, \qquad i = 0, 1, 2, \ldots . \tag{5.15}$$

We may justify (5.15) by noting that, except in the presence of degeneracy, the solution of (5.14) satisfies

$$\mathbf{c}_i \equiv \mathbf{v}_{i+1} - \mathbf{v}_i = \sum_{j=1}^M \boldsymbol{\beta}_j \gamma_j^i . \tag{5.16}$$

$\{\mathbf{v}_i, i = 0, 1, 2, \ldots\}$ is a sequence of partial sums of a vector-valued geometric series generated by M components; it is convergent if each $|\gamma_j| < 1$. In the scalar case, theory based on de Montessus's theorem shows that $[L/M]$ Padé approximants are exact, and consequently also $\varepsilon_{2M}^{(i)}, i = 0, 1, 2, \ldots$. The proof of (5.15) requires the construction of an isomorphism between the complex-valued vectors $\mathbf{v}_i$ and a Clifford algebra [McCleod, 1971].

The approach to matrix Padé approximants adopted in this section follows that of Bessis [1973]. For other approaches and selected applications, we refer to Wynn [1962b, 1963, 1964], Bessis et al. [1974], Hofman et al. [1976], Starkand [1976], Brezinski [1977], and Pindor [1979b].

1.6 Padé–Tchebycheff and Padé–Fourier Approximants etc.

A formal series of Tchebycheff polynomials

$$f(z) = \sideset{}{'}\sum_{i=0}^{\infty} c_i T_i(z) = \tfrac{1}{2} c_0 + c_1 T_1(z) + c_2 T_2(z) + \cdots \tag{6.1}$$

is given, where Σ' denotes that the first coefficient c_0 is to be replaced by $c_0/2$. The Padé–Tchebycheff problem is the construction of an $[L/M]$-type rational approximant using only the first $L+M+1$ terms of the series (6.1). It may well be the case that a Baker–Gammel approximant is more suitable, as described in Section 1.2. In this section, we consider rational approximants exclusively. We seek an approximant of the type

$$\frac{A(z)}{B(z)} = \frac{\sideset{}{'}\sum_{i=0}^{L} a_i T_i(z)}{\sideset{}{'}\sum_{j=0}^{M} b_j T_j(z)} \tag{6.2}$$

which approximates $f(z)$ in (6.1). We proceed by analogy with the methods of Part I, Chapter 1, and treat the difficulties as they occur. The properties of Tchebycheff polynomials needed here follow from the definition

$$T_m(z) = \cos(m \cos^{-1} z) \tag{6.3}$$

The recurrence relation is

$$T_{m+1}(z) - 2z T_m(z) + T_{m-1}(z) = 0, \qquad m = 1, 2, \dots, \tag{6.4}$$

which is initialized by $T_0(z) = 1$, $T_1(z) = z$. The multiplication law is

$$T_i(z) T_j(z) = \tfrac{1}{2} \left[T_{i+j}(z) + T_{|i-j|}(z) \right]. \tag{6.5}$$

The orthogonality property is

$$\int_{-1}^{1} T_i(x) T_j(x) \frac{dx}{\sqrt{1-x^2}} = \frac{\pi}{2} (\delta_{ij} + \delta_{i0} \delta_{0j}). \tag{6.6}$$

Equations (6.4)–(6.6) follow immediately from (6.3). To effect the approximation, we require

$$\sum_{i=0}^{\infty}{}' c_i T_i(z) \approx \frac{\displaystyle\sum_{i=0}^{L}{}' a_i T_i(z)}{\displaystyle\sum_{j=0}^{M}{}' b_j T_j(z)}. \tag{6.7}$$

To give meaning to this approximate equality, we cross-multiply and use (6.5) to derive

$$\left[\sum_{j=0}^{M}{}' b_j T_j(z) \right] \left[\sum_{i=0}^{\infty}{}' c_i T_i(z) \right] = \tfrac{1}{2} \sum_{i=0}^{\infty}{}' \left[\sum_{j=0}^{M}{}' b_j (c_{i+j} + c_{|i-j|}) \right] T_i(z)$$

$$\approx \sum_{i=0}^{L}{}' a_i T_i(z). \tag{6.8}$$

We interpret the latter approximate equality by equating coefficients of $T_i(z)$ for $i = 0, 1, \dots, L+M$, yielding

$$\tfrac{1}{2} \sum_{j=0}^{M}{}' b_j (c_{i+j} + c_{|i-j|}) = 0, \qquad i = L+1, \dots, L+M \tag{6.9a}$$

$$\tfrac{1}{2} \sum_{j=0}^{M}{}' b_j (c_{i+j} + c_{|i-j|}) = a_i, \qquad i = 0, 1, \dots, L. \tag{6.9b}$$

The equations (6.9a) determine $\{b_j\}$ using the given coefficients, and then the equations (6.9b) determine $\{a_i\}$. This appears to be a satisfactory way of calculating the approximants (6.2) until one notices that the data coefficients occurring in (6.9) are $c_0, c_1, \ldots, c_{L+2M}$. An unfortunate consequence of the multiplication law (6.5) is that the formation of an $[L/M]$ approximant needs $L+2M+1$ coefficients of the given series. Obviously, this approximation scheme is uneconomic. Padé–Tchebycheff approximants defined in this way [Maehly, 1956; Holdeman, 1969] are sometimes called "cross-multiplied" approximants [Fleischer, 1973b], to emphasize their derivation from (6.8). The simplest, but rather unsatisfactory solution to the dilemma which occurs when only $c_0, c_1, \ldots, c_{L+M}$ are specified is to use (6.9) with $c_{L+M+1}, c_{L+M+2}, \ldots, c_{L+2M}$ reset to zero.

To avoid these difficulties, one may revert to (6.2) and expand the denominator using

$$\left[\sum_{j=0}^{M} {}' b_j T_j(z) \right]^{-1} = \sum_{j=0}^{\infty} {}' \beta_j T_j(z), \qquad (6.10)$$

where the coefficients β_i are given by using (6.6) as

$$\beta_i = \sqrt{\frac{2}{\pi}} \int_{-1}^{1} \frac{T_i(x)}{\displaystyle\sum_{j=0}^{M} {}' b_j T_j(x)} \frac{dx}{\sqrt{1-x^2}}, \qquad i = 0, 1, 2, \ldots. \qquad (6.11)$$

Substituting (6.10) into (6.2) and equating coefficients of $T_i(x)$, we find that

$$\frac{1}{2} \sum_{j=0}^{L} {}' a_j \left(\beta_{i+j} + \beta_{|i-j|} \right) = c_i, \qquad i = 0, 1, \ldots, L+M.$$

Substituting from (6.11) for β_{i+j} and $\beta_{|i-j|}$, we obtain a horrific system of nonlinear equations for $a_0, \ldots, a_L, b_1, \ldots, b_M$. The approximants so defined are called "properly expanded" approximants [Fleisher, 1973b] and are seldom used.

The most satisfactory approach, both aesthetically and numerically, is due to Clenshaw and Lord [1974]. Using the principles of Baker–Gammel approximants, we assume that the data coefficients $c_0, c_1, \ldots, c_{L+M}$ in (6.1) originate from the expansion of a rational fraction of type $[L/M]$. The $\{c_j\}$ may then be expressed as

$$c_j = \sum_{k=1}^{M} \alpha_k t_k^j \qquad \text{for all} \quad j \geq \max(L-M+1, 0). \qquad (6.12)$$

Consequently $\{c_j\}$ satisfy an $(M+1)$-term recurrence relation

$$\sum_{j=0}^{M} \gamma_j c_{|k-j|} = 0, \qquad k = L+1, L+2, \ldots, \tag{6.13}$$

and each t_k is a root of the polynomial

$$\sum_{j=0}^{M} \gamma_j t^j = 0. \tag{6.14}$$

We assume that $c_j \to 0$ as $j \to \infty$, as is normally the case for Tchebycheff series, and so $|t_k| < 1$. Using (6.12), the equations (6.9a) for the coefficients b_j become

$$\sum_{j=0}^{M}{}' b_j \sum_{k=1}^{M} \alpha_k \left(t_k^{i+j} + t_k^{i-j} \right) = 0, \qquad i = L+1, L+2, \ldots, L+M,$$

i.e.

$$\sum_{k=1}^{M} \alpha_k t_k^i \sum_{j=0}^{M}{}' b_j \left(t_k^j + t_k^{-j} \right) = 0, \qquad i = L+1, L+2, \ldots, L+M. \tag{6.15}$$

We next prove a theorem about the denominator $B(z)$.

THEOREM 1.6.1. *Let* $z = \frac{1}{2}(t + t^{-1})$ *and* $\beta(t) = B(z)$. *Then, with the definitions* (6.12)–(6.15),

$$\beta(t) = \mu \left[\sum_{j=0}^{M} \gamma_j t^j \right] \left[\sum_{j=0}^{M} \gamma_j t^{-j} \right]. \tag{6.16}$$

Proof. $B(z)$ is a polynomial of order M having roots $z_1, z_2, \ldots, z_M$. Let $z_k = \frac{1}{2}(\tau_k + \tau_k^{-1})$, and take $|\tau_k| \leqslant 1$, since both τ_k and τ_k^{-1} correspond to the same z_k. Therefore

$$\beta(t) = \mu_1 t^{-M} \prod_{i=1}^{M} (t - \tau_i) \prod_{i=1}^{M} \left(t - \tau_i^{-1} \right),$$

where μ_1 [like μ in (6.16)] is just a normalizing constant. By definition,

$$B(z) = \sum_{j=0}^{M}{}' b_j T_j(z),$$

and therefore, by the definition of z in terms of t and by (6.3),

$$\beta(t) = \tfrac{1}{2} \sum' b_j(t^j + t^{-j}).$$ (6.17)

Consequently (6.15) is satisfied if t_k is a root of (6.17) and $|t_k| < 1$, and so we may take $t_k = \tau_k$ for each root of (6.17). Hence

$$\sum_{j=0}^{M} \gamma_j t^j \propto \prod_{i=0}^{M} (t - \tau_i)$$

and (6.16) is established.

The algorithm of Clenshaw and Lord consists of determining $\{\gamma_j\}$ from (6.13), taking $\gamma_0 = 1$. From (6.16), the denominator coefficients are given by

$$q_j = \mu \sum_{i=0}^{M-j} \gamma_i \gamma_{j+i}, \qquad j = 0, 1, 2, \ldots, M,$$

which we may normalize by choosing μ such that $q_0 = 2$. The numerator coefficients follow from (6.9b), and the algorithm is specified completely.

The method works best if the numbers $|c_j|$, $j > L + M$, "predicted" by (6.13) tend to zero rapidly. In this case, the zeros of $\beta(t)$ are far from the perimeter of the unit circle of the t-plane, which corresponds to the real interval $-1 \leqslant z \leqslant 1$ in the z-plane. In fact, the capacity of the Clenshaw–Lord algorithm to give near-best rational approximants on $[-1, 1]$ to a variety of test functions is significant and useful. An alternative to the Clenshaw–Lord rational approximants for Tchebycheff series is the scheme of Gragg and Johnson [1974]. With the same conventions as earlier in this section, we assume that the Tchebycheff series

$$f(z) = \frac{c_0}{2} + c_1 T_1(z) + c_2 T_2(z) + \cdots$$

is given; we define $z = (t + t^{-1})/2$,

$$f(z) = \psi(t) = \tfrac{1}{2} \sum_{n=-\infty}^{\infty} c_{|n|} t^n,$$ (6.18)

and also

$$\phi(t) = \tfrac{1}{2} \sum_{n=0}^{\infty}{}' c_n t^n.$$ (6.19)

Thus we can reconstruct $f(z)$ from $\phi(t)$ as

$$f(z) = \phi(t) + \phi(t^{-1}).$$ (6.20)

In order to form an $[L/M]$-type Padé–Tchebycheff rational approximant of $f(z)$, we first construct the $[L/M]$ Padé approximant of $\phi(t)$. Let

$$[L/M]_\phi(t)=\frac{\displaystyle\sum_{i=0}^{L}\alpha_i t^i}{\displaystyle\sum_{j=0}^{M}\beta_j t^j}. \tag{6.21}$$

With (6.20) in mind, we then construct

$$r^{[L/M]}(z)=[L/M]_\phi(t)+[L/M]_\phi(t^{-1})$$

$$=\frac{\displaystyle\sum_{i=0}^{L}\alpha_i t^i\sum_{j=0}^{M}\beta_j t^{-j}+\sum_{i=0}^{L}\alpha_i t^{-i}\sum_{j=0}^{M}\beta_j t^j}{\displaystyle\sum_{i=0}^{M}\sum_{j=0}^{M}\beta_i\beta_j t^{j-i}}$$

$$=\frac{\displaystyle\sum_{i=0}^{L}\sum_{j=0}^{M}\alpha_i\beta_j(t^{i-j}+t^{j-i})}{\displaystyle\sum_{i=0}^{M}\beta_i^2+\sum_{i=1}^{M}\sum_{j=0}^{i-1}\beta_i\beta_j(t^{i-j}+t^{j-i})}$$

$$=\frac{\displaystyle 2\sum_{i=0}^{L}\sum_{j=0}^{M}\alpha_i\beta_j T_{|i-j|}(z)}{\displaystyle\sum_{i=0}^{M}\beta_i^2+2\sum_{i=1}^{M}\sum_{j=0}^{i-1}\beta_i\beta_j T_{i-j}(z)}$$

$$=\frac{\displaystyle\sum_{i=0}^{L}\sum_{j=0}^{M}\alpha_i\beta_j T_{|i-j|}(z)}{\displaystyle 2\sum_{k=0}^{M}{}'\, T_k(z)\left\{\sum_{i=k}^{M}\beta_i\beta_{i-k}\right\}}. \tag{6.22}$$

If $L\geqslant M$, we see that $r^{[L/M]}(z)$ defined by (6.22) is a rational fraction of type $[L/M]$. Hence Equations (6.19), (6.21), and (6.22) formally define a Padé–Tchebycheff approximant $r^{[L/M]}(z)$ to a given Tchebycheff series when $L\geqslant M$.

Next we consider some other types of series. The "cross-multiplied" and "properly expanded" Padé–Tchebycheff approximation schemes discussed earlier can easily be modified to apply to other orthogonal polynomial series, in view of the multiplication law

$$p_i(z)p_j(z)= \sum_{k=0}^{i+j} d_k^{(i,j)}p_k(z)$$

for orthogonal polynomials. The method of Gragg and Johnson applies naturally to Fourier series and to Laurent series as follows:

FOURIER–PADÉ APPROXIMANTS ETC.

$$F(\theta)= \sum_{k=0}^{\infty}{}' \, (a_k\cos k\theta+b_k\sin k\theta) \tag{6.23}$$

is a given formal trigonometric series with real coefficients a_k, b_k. A rational approximant to $F(\theta)$ is required. A simple approach is to take $c_k=a_k-ib_k$ and $z=e^{i\theta}$, reducing (6.18) to

$$F(\theta)=f(z)=\mathrm{Re}\sum_{k=0}^{\infty} c_k z^k. \tag{6.24}$$

A rational approximant to $F(\theta)$ is defined by

$$F^{\mathrm{FP}}(\theta)=\mathrm{Re}[L/M]_f(z)$$

$$=\frac{1}{2}\left(\frac{P^{[L/M]}(z)}{Q^{[L/M]}(z)} + \frac{P^{[L/M]}(z)^*}{Q^{[L/M]}(z)^*} \right).$$

Writing $P^{[L/M]}(z)=\sum_{j=0}^{L}p_j z^j$, $Q^{[L/M]}(z)=\sum_{j=0}^{M}q_j z^j$, and $z=e^{i\theta}$,

$$F^{\mathrm{FP}}(\theta)=\frac{\displaystyle\sum_{j=0}^{L}\sum_{k=0}^{M}\left\{\mathrm{Re}(p_j q_k^*)\cos(j-k)\theta-\mathrm{Im}(p_j q_k^*)\sin(j-k)\theta\right\}}{\displaystyle\sum_{j=0}^{M}\sum_{k=0}^{M}\left\{\mathrm{Re}(q_j q_k^*)\cos(j-k)\theta-\mathrm{Im}(q_j q_k^*)\sin(j-k)\theta\right\}}.$$

This is just one of several possible ways of forming a rational approximant for the Fourier series (6.23). Equation (6.24) can be reexpressed as

$$F(\theta)=f(z)= \sum_{k=-\infty}^{\infty}{}' \, c_k z^k \qquad \text{with} \quad c_{-k}=c_k^*,$$

introducing a Laurent series. There are a host of different approaches to the

Padé–Laurent problem. Construction of approximants with attractive mathematical properties and convergence theorems is still at an early stage for Padé–Laurent series, and much remains to be done.

For further details, we refer to Cheney [1966], Fleischer [1972, 1973a], Gragg [1977], and Chisholm and Common [1980].

Exercise 1. By considering the generating function

$$\frac{1-hz}{1-2hz-h^2} = \sum_{n=0}^{\infty} h^n T_n(z),$$

show that the algorithm of Clenshaw and Lord generates Baker–Gammel approximants to the series (6.1) for $L \geq M-1$.

Exercise 2. Is it a good idea to use (6.13) as a "predictor" of c_j for $j \geq L+M$ to test whether the predicted $|c_j| \to 0$?

Exercise 3. Verify (6.8).

Exercise 4. Investigate the connection between the Gragg–Johnson scheme (6.18)–(6.24) and the formation of Baker–Gammel approximants for $L \geq M$.

CHAPTER 2

Connection with
Integral Equations
and Quantum Mechanics

2.1 The General Method and Finite-Rank Kernels

We will consider linear, inhomogeneous integral equations of the type

$$f(x) = g(x) + \lambda \int_a^b A(x, y) f(y) \, dy. \tag{1.1}$$

These are sometimes called Fredholm equations of the second kind. If (1.1) is solved iteratively, the Liouville–Neumann expansion results:

$$f(x) = g(x) + \lambda \int_a^b A(x, y) g(y) \, dy$$
$$+ \lambda^2 \int_a^b A(x, y_1) \int_a^b A(y_1, y) g(y) \, dy \, dy_1 + \cdots. \tag{1.2}$$

The expansion (1.2) formally satisfies (1.1), and is a solution of (1.1) if the series (1.2) is uniformly convergent. It is usually the case that the quantity of interest is not $f(x)$ but rather

$$\langle h | f \rangle \equiv \int_a^b h^*(x) f(x) \, dx, \tag{1.3}$$

ENCYCLOPEDIA OF MATHEMATICS and Its Applications, Gian-Carlo Rota (ed.). Vol. 14: George A. Baker, Jr., and Peter R. Graves-Morris, Padé Approximants, Part II: Extensions and Applications   ISBN 0-201-13513-2

and then

$$\langle h|f\rangle = \int_a^b h^*(x)g(x)\,dx + \lambda \int_a^b \int_a^b h^*(x)A(x,y)g(y)\,dx\,dy$$

$$+\lambda^2 \int_a^b \int_a^b \int_a^b h^*(x)A(x,y_1)A(y_1,y)g(y)\,dx\,dy_1\,dy + \cdots \quad (1.4)$$

for suitable convergence conditions. We defer detailed questions of convergence to Section 2.2. The series (1.4) is a power series in λ, which we write concisely as

$$\langle h|f\rangle = c_0 + \lambda c_1 + \lambda^2 c_2 + \cdots, \quad (1.5)$$

where

$$c_n = \int_a^b \int_a^b \cdots \int_a^b h^*(y_1)A(y_1,y_2)A(y_2,y_3)\cdots A(y_{n-1},y_n)g(y_n)\,dy_1\cdots dy_n.$$

$$(1.6)$$

Provided all the coefficients $\{c_n\}$ of (1.6) exist, we may sum the series (1.5) using Padé approximants in the hope that the method converges. The basic application of Padé approximants in integral equations is very simple: use the iterative solution, and use Padé approximants to interpret it. In this chapter we will consider a variety of applications. In some instances the Padé method may be proved to converge, and we know of no circumstances where the Padé method, properly applied, has failed for integral equations in genuine applications.

We will now consider an important and very particular class of kernels which are called finite-rank kernels. They may be written as

$$A(x,y) = \sum_{j=1}^n X_j(x)Y_j(y). \quad (1.7)$$

We will suppose that n is the smallest integer allowing the representation (1.7), so as to obviate unnecessary later difficulties. Finite-rank kernels are important because they allow the basic integral equation (1.1) to be solved exactly, and this solution sheds a great deal of light on the nature of the solution with more general kernels.

We will find the explicit solution of (1.1) using the finite rank kernel (1.6) and show that the Padé approximant solution is exact.

Substitute (1.7) in (1.1) to obtain

$$f(x) = g(x) + \lambda \sum_{j=1}^n X_j(x) \int_a^b Y_j(y)f(y)\,dy$$

$$= g(x) + \lambda \sum_{j=1}^n f_j X_j(x), \quad (1.8)$$

where

$$f_j = \int_a^b Y_j(y)f(y)\,dy, \quad j=1,2,\ldots,n. \quad (1.9)$$

Let

$$g_i = \int_a^b Y_i(x)g(x)\,dx, \qquad i=1,2,\ldots,n, \qquad (1.10)$$

$$h_i = \int_a^b X_i(x)h^*(x)\,dx, \qquad i=1,2,\ldots,n, \qquad (1.11)$$

and

$$d_{ij} = \int_a^b Y_i(y)X_j(y)\,dy, \qquad i,j=1,2,\ldots,n. \qquad (1.12)$$

Equations (1.9)–(1.12) define the basic vectors $\mathbf{f}$, $\mathbf{g}$, and $\mathbf{h}$ and the matrix D which are needed presently. By integrating (1.8) over $Y_i(x)$, we find

$$f_i = g_i + \lambda \sum_{j=1}^n d_{ij}f_j,$$

i.e.,

$$\mathbf{f} = \mathbf{g} + \lambda D\mathbf{f},$$

and so $\mathbf{f} = (I-\lambda D)^{-1}\mathbf{g}$, provided $I-\lambda D$ is invertible. By substituting this formula for f_i in (1.8), we find

$$f(x) = g(x) + \lambda \sum_{i=1}^n \left[(I-\lambda D)^{-1}\mathbf{g}\right]_i X_i(x), \qquad (1.13)$$

and so

$$\langle h|f\rangle = \int_a^b h^*(x)g(x)\,dx + \lambda \sum_{i=1}^n \sum_{j=1}^n h_i\left[(I-\lambda D)^{-1}\right]_{ij}g_j. \qquad (1.14)$$

Hence (1.14) provides a solution of the problem, using the definitions (1.10), (1.11), and (1.12). We notice that the inverse matrix $(I-\lambda D)^{-1}$ is assumed to exist (and it certainly does if λ is sufficiently small), then the power-series expansion of (1.14) is valid. In general, we notice that the solution (1.14) is a rational function of λ. We demonstrate this by writing

$$(I-\lambda D)^{-1} = \frac{\mathrm{adj}(I-\lambda D)}{\det(I-\lambda D)},$$

where the adjugate matrix, adj A, is defined to be the transposed matrix of cofactors of A. Hence we see that (1.14) implies that

$$\langle h | f \rangle = \frac{N(\lambda)}{D(\lambda)}, \tag{1.15}$$

where $N(\lambda)$ and $D(\lambda)$ are normally polynomials of degree n. In particular, any $[L/M]$ Padé approximant to the power series of (1.14) with $L \geqslant n$ and $M \geqslant n$ is exact, from Theorem I.1.4.4 [Chisholm, 1963]. We are also naturally led to consider diagonal Padé approximants for the solution of integral equations.

An important and useful conclusion to this section is that the diagonal Padé approximants to the solution $\langle h | f \rangle$ of the integral equation (1.1) converge to the exact solution, if the kernel is of finite rank.

More generally, for any $\alpha \in (0, 1)$, the $[L/M]$ Padé approximants to the expansion of $\langle h | f \rangle$, (1.4), converge to the solution $\langle h | f \rangle$ of the integral equation (1.1) with a finite rank kernel, provided

$$\alpha < L/M < \alpha^{-1} \quad \text{and} \quad \min(L, M) \to \infty.$$

Furthermore, for a kernel of rank n, the $[L/M]$ Padé approximants are exact provided $\min(L, M) \geqslant n$.

We defer questions of convergence to the next section; however, the answers to such questions with finite-rank kernels are scarcely difficult, given the existence of an explicit analytical solution.

Exercise 1. Solve the integral equation $f(x) = g(x) + \lambda \int_0^1 (1 + xy) f(y) \, dy$

  (i) by the method of Equation (1.8);
  (ii) by using Padé approximants.

Exercise 2. Solve the integral equation (1.1) with $A(x, y) = 1$. Note the apparent discrepancy with (1.15) regarding $\deg\{N(\lambda)\}$.

2.2 Padé Approximants and Integral Equations with Compact Kernels

In the previous section, we stated the Padé method and motivated it by the study of finite-rank kernels, following the historical approach. In this section we will summarize some of the results of the theory of integral equations and relate these to the theory of Padé approximants. The underlying theme comprises two parts: establishing that $\langle h | f \rangle$ of (1.4) is a meromorphic function of λ, and selecting the sequence of Padé approximants which converge to such a function. If special properties of the integral equation (1.1) imply that $\langle h | f \rangle$ is not merely meromorphic (for example, $\langle h | f \rangle$ might be a Stieltjes series), the methods of Section 2.4 may be applicable, and stronger conclusions then follow. Presently we consider the general integral equation of the type

$$f(x) = g(x) + \lambda \int_a^b A(x, y) f(y) \, dy, \tag{2.1}$$

and its solution in the form

$$\langle h | f \rangle \equiv \int_a^b h^*(x) f(x) \, dx \tag{2.2}$$

as at the start of Section 2. Let us recall the results of Hilbert-space theory. $\mathcal{H}$ is a linear space with elements $f, g, h, \ldots$ [not yet to be associated with f, g, and h of (2.1) and (2.2)] including a null element 0. To every pair of elements there is an inner product $\langle f | g \rangle$ satisfying the usual rules, including $\langle f | g \rangle = \langle g | f \rangle^*$. Each element f has a norm $\| f \| = \langle f | f \rangle^{1/2}$ given by the inner product. Every Cauchy sequence of elements of $\mathcal{H}$ converges to a limit element which belongs to $\mathcal{H}$.

The first important linear space we use is the space $\mathcal{C}[a, b]$ of a continuous functions defined on a closed finite interval $[a, b]$. The inner product is defined by

$$\langle h | g \rangle = \int_a^b h^*(x) g(x) \, dx. \tag{2.3}$$

$\mathcal{C}[a, b]$ is not a Hilbert space, because it is not complete. This means that Cauchy sequences of continuous functions defined on $[a, b]$ may be found which converge to discontinuous functions, such as step functions, using the norm implied by (2.3). We will assume in this context that $g(x)$ of (2.1) belongs to $\mathcal{C}[a, b]$ and that $A(x, y)$ of (2.1) is continuous for all $x, y \in [a, b]$. Then given $\varepsilon > 0$, Weierstrass's approximation theorem states that a polynomial $S_N(x, y)$ of sufficiently high degree N, and depending on ε, exists such that

$$|S_N(x, y) - A(x, y)| < \varepsilon$$

for all $x, y \in [a, b]$. Then $T(x, y)$ is defined by

$$A(x, y) = S_N(x, y) + T(x, y) \tag{2.4}$$

with the property that $\| T \|_\infty < \varepsilon$. This property is sufficient to show that the operator

$$(1 - \lambda T)^{-1} = 1 + \lambda T + \lambda^2 T^2 + \cdots \tag{2.5}$$

exists for all $|\lambda| < \varepsilon^{-1}$. By this we mean that (2.5) converges to a continuous kernel which maps any function from $\mathcal{C}[a, b]$ into a function of $\mathcal{C}[a, b]$. We use (2.4) and (2.5) in (2.1):

$$f = g + \lambda(S_N + T)f, \tag{2.6}$$

$$(1 - \lambda T)f = g + \lambda S_N f,$$

$$f = (1 - \lambda T)^{-1} g + \lambda (1 - \lambda T)^{-1} S_N f. \tag{2.7}$$

Equation (2.7) is an integral equation of finite rank, and so f is meromorphic in λ, with at most N poles in the circle $|\lambda| < \varepsilon^{-1}$. We deduce that $\langle h | f \rangle$ is meromorphic in λ.

The second space we need to consider is the Hilbert space $\mathcal{L}^2(a, b)$, the space of square integrable functions defined on the open interval (a, b). For this space, we allow the possibility that $a = -\infty$, or $b = \infty$, or both. The inner product is the same as (2.3). If a and b are finite, $\mathcal{C}[a, b] \subset \mathcal{L}^2(a, b)$. To make use of this space, we will assume that

$$\int_a^b |g(x)|^2 \, dx < \infty \quad \text{and} \quad \int_a^b \int_a^b |A(x, y)|^2 \, dx \, dy < \infty$$

in this context. There is a theorem [Smithies, 1958, Chapter 3] that, for any $\varepsilon > 0$, any $\mathcal{L}^2$ kernel $A(x, y)$ has an expansion

$$A(x, y) = \sum_{n=1}^{N} \frac{u_n(x) v_n(y)}{\lambda_n} + T(x, y) \tag{2.8}$$

where $\|T\| = \int_a^b \int_a^b |T(x, y)|^2 \, dx \, dy < \varepsilon$. The functions u and v belong to $\mathcal{L}^2(a, b)$ and are normalized eigenfunctions of $AA^\dagger$ and $A^\dagger A$ respectively. An analysis identical to that following (2.6) shows again that $\langle h | f \rangle$ of (2.2) is meromorphic in λ.

We have considered kernels in two different Hilbert spaces, and used them as operators in the sense that

$$Af = \int_a^b A(x, y) f(y) \, dy. \tag{2.9}$$

Further, they are compact operators. An operator is defined to be compact (sometimes called completely continuous) if it transforms every bounded set of elements in the space into a sequence which contains a convergent subsequence. For example, we see that the identity operator and also v defined by $vg = v(x)g(x)$ with $v \not\equiv 0$ are not compact operators. The concept of a compact operator is useful in both Hilbert- and Banach-space theory. A Banach space has all the axioms of a Hilbert space except that the norm is not required to be an inner product. In an arbitrary Banach space, it is not true that every compact operator can be approximated in norm by a kernel of finite rank, so that an equation such as (2.8) is unavailable. But it is true that the resolvent of every compact operator on $\mathcal{B}$ is meromorphic in λ. Referring to (2.1) and (2.2), if $g \in \mathcal{B}$, $h \in \mathcal{B}^*$ (the dual of $\mathcal{B}$), and A is a compact operator on $\mathcal{B}$, then $\langle h | -$ is a meromorphic function of λ.

Three approaches of increasing sophistication have been given which suffice to prove that $\langle h | f \rangle$ is a meromorphic function of λ. Let us turn to consider the theorems and conjectures available to utilize this information.

If a kernel S_N of finite rank N exists which is a good approximation to A, one expects the $[N/N]$ Padé approximant to $\langle h | f \rangle$ to be a good approximation, and this is usually the case. Motivated by the analysis of the previous section, the folklore is that use of the diagonal sequence of Padé approximants for $\langle h | f \rangle$ is an efficient way of solving integral equations occurring in practice. The fact that the method works well in practice is understood as implying that ordinary kernels have (whether or not one can find them) accurate finite-rank approximations in the sense of the relevant norm.

Counterexample [Graves-Morris, 1978b].   There exist continuous functions $g(x)$, $h(x)$ and a continuous kernel $A(x, y)$ for which the sequence of diagonal approximants of (1.4) does not converge.

We use Legendre polynomials $P_n(x)$ to define $\{ p_n(x) = (n + \frac{1}{2})^{1/2} P_n(x),$ $n = 0, 1, 2, \ldots \}$, which is an orthogonal sequence on $[-1, 1]$. We define $g(x) = p_0(x) = 1/\sqrt{2}$ and

$$A(x, y) = \sum_{n=0}^{\infty} \frac{p_{n+1}(x) p_n(y)}{(n+1)^3}. \tag{2.10}$$

Because $|P_n(x)| \leqslant 1$ on $-1 \leqslant x \leqslant 1$, the sum in (2.10) is uniformly convergent (by the Weierstrass M-test) and so $A(x, y)$ is continuous on $-1 \leqslant x \leqslant 1$, $-1 \leqslant y \leqslant 1$. Hence the kernel (2.10) is compact on the Hilbert space $\mathcal{L}^2(-1, 1)$.

We define $h(x)$ by

$$h(x) = \sum_{n=0}^{\infty} h_n p_n(x) = \sum_{n=0}^{\infty} \{ n! \}^3 c_n p_n(x), \tag{2.11}$$

where the coefficients c_n are specified as follows. Indices n_k are defined iteratively by

$$n_1 = 1, \qquad n_{k+1} = 2n_k + 1;$$

we define $c_n = \alpha_k \lambda_k^{-1}$ if n is such that $n_k \leq n < n_{k+1}$; α_k is defined by

$$\alpha_k = \{(2n)!\}^{-4} \min\{|\lambda_k|^{n_k}, |\lambda_k|^{2n_k}\}, \qquad (2.12)$$

and the sequence $\{\lambda_k\}$ is defined to be any set of points (allowing repetition) which is dense in the λ-plane. The choice (2.12) of α_k ensures that

$$|c_n| < (n!)^{-4}, \qquad n = 1, 2, \ldots,$$

so that (2.11) is uniformly convergent on $-1 \leq x \leq 1$. We have constructed $g(x)$, $A(x, y)$, and $h(x)$ specifically so that the series

$$\langle h | f \rangle = \sum_{n=0}^{\infty} \lambda^n \langle h | A^n | g \rangle = \sum_{n=0}^{\infty} c_n \lambda^n$$

is essentially the same as the series constructed by Gammel (see Part I, Section 6.7), the only, minor difference being an extra factor $(2n_k)!^{-3}$ in (2.12) which is not in Gammel's example, but which is needed for reasons associated with the convergence of (2.10) and (2.11). Using Gammel's argument, it follows that the sequence of diagonal approximants of (1.4) diverges on a dense set of points in the whole λ-plane. This establishes the counterexample.

The conjecture of Baker, Gammel and Wills [1961], in this context, is that at least an infinite subsequence of diagonal Padé approximants to any meromorphic function converge in a bounded region of the λ-plane, except on arbitrary open sets enclosing the poles of the function. There is as yet, no proof of this conjecture, but there are strong indications of its validity.

The theorem of Nuttall [1970b] of convergence in measure is a solid result, albeit not quite the result one might have expected. In this context, the theorem states that paradiagonal $[M+J/M]$ sequences converge to $\langle h | f \rangle$ in measure. This means, colloquially, that the area of the region of the complex λ-plane where any approximant of sufficiently high order is less accurate than a given error bound is small. Of course, the location of this region is unknown, and it depends on M. Nuttall's theorem is strengthened by Pommerenke's modification [1973], the relevant portion of which replaces the notion of measure by capacity. This change has the effect of restricting the maximum diameter of any closed connected exceptional region of the λ-plane instead of restricting its measure. This result and the restrictions on the size of the exceptional region are explained in Part I, Section 6.6.

To summarize, we do not yet know the conditions ensuring that the general method of Sections 2.1 and 2.2 with a paradiagonal sequence of Padé approximants converges, but we have every reason to believe that the "bad" approximants of any sequence are rare.

Lastly, we consider the poles of the inverse operator $(1-\lambda A)^{-1}$. Clearly, the Padé approximants of $\langle h|(1-\lambda A)^{-1}|g\rangle$ do not converge at the poles in the usual way. One can introduce the chordal metric (convergence on the sphere) as described in Part I, Section 6.4 to include this case, and this is often desirable. If inspection of a paradiagonal sequence of Padé approximants reveals a stable, convergent sequence of poles, in practice these are always poles of the solution $\langle h| f \rangle$. Location of these poles enables us to find the values of λ which solve the homogeneous integral equation

$$f(x)=\lambda\int_a^b A(x, y)f(y)\,dy. \qquad (2.13)$$

The analysis of (2.6) and (2.7), showing that (2.1) has a unique solution, except at $\{\lambda=\lambda_n, n=1,2,3,\dots\}$, with

$$|\lambda_n|\geqslant\|A(x, y)\|^{-1},$$

implies that (2.13) has no solution except possibly at $\{\lambda=\lambda_n, n=1,2,3,\dots\}$. The Padé method of finding these values is to solve the inhomogeneous equation (2.1) with some arbitrary g, locate the poles of the paradiagonal sequence of approximants, and extrapolate their limiting points. In principle, g must be chosen so that it is not orthogonal to the relevant eigenfunction. It should also be remembered that operators of the type $(1-\lambda A)^{-1}$ discussed in this and the previous section need not have any poles at all. For example, the integral equation

$$f(x)=g(x)+\lambda\int_a^x A(x, y)f(y)\,dy$$

is a Volterra integral equation, and if A is an $\mathcal{L}^2$ kernel, $\langle h| f \rangle$ is an entire function of λ. This statement means that its Maclaurin series converges for any value of λ. In such a case, one would not use the Padé method except for the explicit purpose of accelerating convergence.

The books by Smithies [1958], Tricomi [1957], and Riesz and Nagy [1955] are good general texts on integral equations. For further details of the connection with Padé approximants, we refer to Baker and Gammel [1961], Vorobyev [1965], and Graves-Morris [1973].

2.3 Projection Techniques

We discuss in this section the solution of the standard inhomogeneous linear integral equation which takes the form

$$f(x)=g(x)+\lambda\int_a^b A(x, y)f(y)\,dy. \qquad (3.1)$$

It will be convenient to write (3.1) in the Hilbert-space formalism using the Dirac notation as

$$|f\rangle = |g\rangle + \lambda A |f\rangle. \tag{3.2}$$

We have in mind the Hilbert spaces of $\mathfrak{L}^2$ functions on either finite or infinite ranges (a, b), and A is a linear operator on the Hilbert space. The theory we discuss is quite general, and applications in Banach spaces are sometimes appropriate. The basic problem we wish to treat is the evaluation of the overlap integral

$$\langle h | f \rangle = \int_a^b h^*(x) f(x)\, dx,$$

where $\langle h |$ is another vector from the Hilbert space, and $| f \rangle$ is the solution of (3.1, 3.2). By expanding the iterative solution of (3.1, 3.2), we find

$$|f\rangle = |g\rangle + \lambda A |g\rangle + \lambda^2 A^2 |g\rangle + \cdots, \tag{3.3}$$

and therefore

$$\langle h | f \rangle = \langle h | g \rangle + \lambda \langle h | A | g \rangle + \lambda^2 \langle h | A^2 | g \rangle + \cdots \tag{3.4}$$

$$= c_0 + c_1 \lambda + c_2 \lambda^2 + \cdots, \tag{3.5}$$

where we have defined $c_i = \langle h | A^i | g \rangle$, $i = 0, 1, 2, \ldots$. The equalities are well defined if $|\lambda| \cdot \|A\| < 1$, and they are only formal if $|\lambda| \cdot \|A\| \geqslant 1$. It is clear from (3.4) that the basic vectors $\langle h |$ and $| g \rangle$ enter the problem more symmetrically than the original formulation suggests.

If $A(x, y)$ is a constant, (3.4) becomes a geometric series,

$$\langle h | f \rangle = \frac{c_0}{1 - \lambda A},$$

and the $[0/1]$ Padé approximant is exact . More generally, for the $[N-1/N]$ Padé approximant, it follows from (I.1.3.6) that

$$[N-1/N]_{\langle h | f \rangle} = \sum_{i,j=0}^{N-1} c_i [C^{-1}]_{ij} c_j, \tag{3.6}$$

where the matrix

$$C_{ij} \equiv c_{i+j} - \lambda c_{i+j+1}. \tag{3.7}$$

We will see in this section that we may interpret the $[N-1/N]$ Padé approximants to the Neumann series as the exact solution of an integral equation similar to the original equation (3.1), (3.2).

Let us consider the subspace of the original Hilbert space spanned by the N elements

$$|g\rangle, \ A|g\rangle, \ A^2|g\rangle, ..., \ A^{N-1}|g\rangle$$

and define

$$\left.\begin{array}{l}|\psi_i\rangle = A^{i-1}|g\rangle \\ \langle\psi_i'| = \langle h|A^{i-1}\end{array}\right\} \qquad i = 1, 2, ..., N. \tag{3.8}$$

Then we define S_N to be the space spanned by $|\psi_1\rangle, |\psi_2\rangle, ..., |\psi_N\rangle$, and S_N' to be the space spanned by $\langle\psi_1'|, \langle\psi_2'|, ..., \langle\psi_N'|$. We define the matrix R by

$$R_{ij} = \langle\psi_i'|\psi_j\rangle = c_{i+j-2}, \qquad i, j = 1, 2, ..., \tag{3.9}$$

and the projection operator $\mathcal{P}_N$ by

$$\mathcal{P}_N = \sum_{i,j=1}^{N} |\psi_j\rangle (R^{-1})_{ij} \langle\psi_i'|. \tag{3.10}$$

It is immediately clear by matrix multiplication that

$$\mathcal{P}_N^2 = \mathcal{P}_N$$

so that $\mathcal{P}_N$ is a projection operator. $\mathcal{P}_N$ projects Hilbert-space vectors into the space S_N. We see that

$$\mathcal{P}_N|\psi_i\rangle = |\psi_i\rangle \qquad \text{for} \quad i = 1, 2, ..., N,$$

and so $\mathcal{P}_N$ is represented by the identity on S_N. We define T_N to be the complement of S_N in the Hilbert space. In the general case, T_N contains $A^N|g\rangle$, $A^{N+1}|g\rangle$, and many other vectors. These are mapped by $\mathcal{P}_N$ into S_N, but unfortunately it is a difficult matter to put a bound on $\|\mathcal{P}_N\|$ in this case. $\mathcal{P}_N$ is not an orthogonal projection, but an oblique projection, and the following analogy may help to understand it. The horizontal plane of a sundial is the space S_N, and the vertical is T_N. The shadow of the gnomon always lies in S_N, and may have any length, depending on the elevation of the sun. In general it is clear that

$$\|\mathcal{P}_N\| \geqslant 1,$$

and $\mathcal{P}_N$ may be represented schematically by

$$\mathcal{P}_N\begin{pmatrix} S_N \\ T_N \end{pmatrix} = \begin{pmatrix} I & X \\ 0 & 0 \end{pmatrix}\begin{pmatrix} S_N \\ T_N \end{pmatrix}.$$

It turns out that the $[N-1/N]$ Padé approximant solution of (3.1), (3.2) is the exact solution of the integral equation

$$|f_N\rangle = |g\rangle + \lambda \mathcal{P}_N A |f_N\rangle. \tag{3.11}$$

Proof. Equation (3.11) implies that $|f_N\rangle \in S_N$, and therefore we may write

$$|f_N\rangle = \sum_{i=1}^{N} d_i |\psi_i\rangle. \tag{3.12}$$

To find the $\{d_i\}$, substitute (3.10) and (3.12) in (3.11) to give

$$\sum_{i=1}^{N} d_i |\psi_i\rangle = |g\rangle + \lambda \sum_{i,j,k=1}^{N} |\psi_i\rangle (R^{-1})_{ij} \langle \psi_j' | A |\psi_k\rangle d_k.$$

Forming the inner product with $\langle \psi_m' |$ for $m = 1, 2, \ldots, N$ yields N linear equations for the $\{d_i\}$,

$$\sum_{i=1}^{N} d_i R_{mi} = \langle \psi_m' | g \rangle + \lambda \sum_{k=1}^{N} \langle \psi_m' | A |\psi_k\rangle d_k.$$

From (3.8), (3.9),

$$\sum_{k=1}^{N} (c_{m+k-2} - \lambda c_{m+k-1}) d_k = c_{m-1},$$

and again from (3.8), (3.12),

$$\langle h | f_N \rangle = \sum_{i=1}^{N} d_i c_{i-1}$$

$$= \sum_{i,j=0}^{N-1} c_i (C^{-1})_{ij} c_j,$$

where the matrix C has elements given by (3.7). This derivation proves that the $[N-1/N]$ Padé approximant to $\langle h | f \rangle$ given by (3.6) is the exact solution of (3.11).

Before passing by, we note that if $\mathcal{P}_N$ were an orthogonal projection, so that $\mathcal{P}_N T_N = 0$, we could prove that $\mathcal{P}_N A \to A$ as $N \to \infty$ and we would deduce that the Padé approximant solution to (3.1) converges. However, $\mathcal{P}_N$ is an oblique projection and we may only deduce convergence of the Padé approximants to (3.1) when constraints are placed on A, $|g\rangle$, and $\langle h|$ so that $\mathcal{P}$ is approximately orthogonal [Baker, 1975].

We now review the previous method of construction of the Padé-approximant solution of (3.1) using the elements of S_N given by (3.8) in the light of Lanczos method of minimal iterations [Lanczos, 1952]. This was originally due to Garibotti and Villani [1969a], who showed that the Padé denominator of $\langle h | f_N \rangle$ is an orthogonal polynomial developed in the theory of the biorthogonal algorithm. This algorithm develops the previous analysis by using bases of vectors in S_N and S_N' which are biorthogonal.

From the vectors

$$|\psi_i\rangle = A^{i-1}|g\rangle \in S_N, \qquad i=1,2,\ldots,N, \qquad (3.13)$$

and

$$|\psi_j'\rangle = (A^\dagger)^{j-1}|h\rangle \in S_N', \qquad j=1,2,\ldots,N,$$

we construct a new set of functions $|\phi_i\rangle, |\phi_j'\rangle$ which satisfy the orthogonality condition that

$$\langle \phi_i' | \phi_j \rangle = \omega_i \delta_{ij}, \qquad i,j=0,1,\ldots,N-1. \qquad (3.14)$$

The $\{\omega_i\}$ are normalization constants only. In the event that $|g\rangle = |h\rangle$ and $A = A^\dagger$ (i.e. $g(x) = \{h(x)\}^*$ and $A(x,y) = \{A(y,x)\}^*$. It follows that S_N and S_N' are essentially the same space, and the analysis is much simpler, leading to a set of orthogonal vectors in S_N (or S_N') in the usual sense. In the general case, the biorthogonal algorithm is most easily developed by defining polynomial operators sequentially by

$$p_0(A)=1, \qquad (3.15a)$$

$$p_1(A)=A-\alpha_0, \qquad (3.15b)$$

$$p_{i+1}(A)=(A-\alpha_i)p_i(A)-\beta_i p_{i-1}(A) \qquad \text{for} \quad i=1,2,\ldots, \quad (3.15c)$$

and the constants α_i, β_i required in (3.15) are chosen so that

$$\alpha_i = \frac{\langle \phi_i' | A | \phi_i \rangle}{\langle \phi_i' | \phi_i \rangle} \quad \text{and} \quad \beta_i = \frac{\langle \phi_i' | \phi_i \rangle}{\langle \phi_{i-1}' | \phi_{i-1} \rangle} = \frac{\omega_i}{\omega_{i-1}}. \qquad (3.16)$$

We may prove by induction that we may take

$$|\phi_i\rangle = p_i(A)|g\rangle$$

and   (3.17)

$$|\phi_i'\rangle = p_i(A^\dagger)|h\rangle$$

for $i=0,1,2,\ldots$ to obey the biorthogonality condition (3.14). The proof is straightforward, and we omit the details.

In the event that A is a finite-rank operator of rank N, the elements $A^N|g\rangle$, $A^{N+1}|g\rangle,\ldots$ are linearly dependent on elements of S_N. From (3.15) we see that the leading term of $p_i(A)$ is A^i, guaranteeing that the space spanned by $|\phi_0\rangle,|\phi_1\rangle,\ldots,|\phi_{N-1}\rangle$ is S_N for all N. Hence we see that if A is a finite-rank operator, then $|\phi_N\rangle$ is a linear combination of $|\phi_0\rangle,\ldots,|\phi_{N-1}\rangle$ and it follows from the biorthogonality condition (3.14) that $|\phi_N\rangle=0$. We conclude that if A has rank N, then $|\phi_N\rangle=0$, and we prove similarly that $\langle\phi'_N|=0$.

We may adopt the attitude in the general case that by truncating the orthogonalization process at stage N, for N large enough, we have taken sufficiently many dominant eigenfunctions, characterized by their relatively large eigenvalues, into account. At any rate, we will formally curtail the orthogonalization process by taking $|\phi_N\rangle=\langle\phi'_N|=0$. This leads to the truncated integral equation

$$|f_N\rangle=|g\rangle+\lambda\mathcal{P}_N A|f_N\rangle, \qquad ((3.11))$$

which we now solve by an expansion

$$|f_N\rangle=\sum_{j=0}^{N-1} e_j|\phi_j\rangle. \qquad (3.18)$$

Substituting (3.18) in (3.11) and taking the inner product with $\langle\phi'_i|$ yields

$$e_i\omega_i=\omega_0\delta_{i0}+\lambda\langle\phi'_i|A\sum_{j=0}^{N-1} e_j|\phi_j\rangle. \qquad (3.19)$$

From (3.15) we find that for $i=1,2,\ldots,N-2$

$$\langle\phi'_i|A=\langle h|p_i(A)A$$
$$=\langle h|p_{i+1}(A)+\alpha_i p_i(A)+\beta_i p_{i-1}(A)$$
$$=\langle\phi'_{i+1}|+\alpha_i\langle\phi'_i|+\beta_i\langle\phi'_{i-1}|,$$

and hence from (3.19)

$$e_i\omega_i=\lambda(e_{i+1}\omega_{i+1}+\alpha_i e_i\omega_i+\beta_i e_{i-1}\omega_{i-1}).$$

Dividing by ω_i and using (3.16),

$$e_i=\lambda(\beta_{i+1}e_{i+1}+\alpha_i e_i+e_{i-1}). \qquad (3.20a)$$

The initial equation for $i=0$ is

$$e_0 = 1 + \lambda(\beta_1 e_1 + \alpha_0 e_0), \tag{3.20b}$$

and the final equation for $i = N-1$ is

$$e_{N-1} = \lambda(\alpha_{N-1} e_{N-1} + e_{N-2}). \tag{3.20c}$$

Equations (3.20a–c) are N equations for N unknowns $e_0, e_1, \ldots, e_{N-1}$. To solve these equations, let us take $\mu = \lambda^{-1}$, so that they become recurrence relations

$$\beta_1 e_1 + (\alpha_0 - \mu) e_0 + \mu = 0, \tag{3.21a}$$

$$\beta_{i+1} e_{i+1} + (\alpha_i - \mu) e_i + e_{i-1} = 0, \qquad i = 1, 2, \ldots, N-2, \tag{3.21b}$$

and

$$(\alpha_{N-1} - \mu) e_{N-1} + e_{N-2} = 0. \tag{3.21c}$$

To make (3.21b) take the form of (3.15c), substitute

$$e_i' = \omega_i e_i \tag{3.22}$$

and multiply (3.21b) by ω_i to give

$$e_{i+1}' + (\alpha_i - \mu) e_i' + e_{i-1}' \beta_i = 0 \tag{3.23a}$$

with the associated equations

$$e_1' + (\alpha_0 - \mu) e_0' + \omega_0 \mu = 0 \tag{3.23b}$$

and

$$(\alpha_{N-1} - \mu) e_{N-1}' + e_{N-2}' \beta_{N-1} = 0. \tag{3.23c}$$

The solution of the recurrence relation (3.23a) with initial conditions $p_0(\mu) = 1$, $p_1(\mu) = \mu - \alpha_0$ are the polynomials $p_i(\mu)$. Let us define polynomials $\tilde{p}_i(\mu)$ to be the solutions of the recurrence (3.23a) with initial conditions $\tilde{p}_0(\mu) = 0$, and $\tilde{p}_1(\mu) = \mu$. Then the solution of (3.23) is a linear combination of these,

$$e_i' = \gamma p_i(\mu) + \delta \tilde{p}_i(\mu), \tag{3.24}$$

provided (3.23b) and (3.23c) are satisfied. These give

$$\delta + \omega_0 = 0$$

and

$$\gamma p_N(\mu) + \delta \tilde{p}_N(\mu) = 0,$$

from which the values of δ and γ follow. Hence (3.24) and (3.22) and (3.18) give $|f_N\rangle$ explicitly, and we find

$$\langle h| f_N \rangle = e_0 \omega_0 = e_0' = \gamma$$
$$= \frac{\tilde{p}_N(\mu)}{p_N(\mu)} \omega_0.$$

The significance of this result is that we see directly that the Padé-approximant solution of the integral equation has poles at the zeros of $p_N(1/\lambda)$ and zeros at the zeros of $\tilde{p}_N(1/\lambda)$. By writing the equations (3.21a, b) in the form

$$\beta_1\left(\frac{e_1}{e_0}\right) + (\alpha_0 - \mu) + \frac{\mu}{e_0} = 0$$

and

$$\beta_{i+1}\left(\frac{e_{i+1}}{e_i}\right) + \alpha_i - \mu + \frac{e_{i-1}}{e_i} = 0,$$

it follows that

$$e_0 = \frac{-\mu}{\alpha_0 - \mu} + \frac{\beta_1}{\alpha_1 - \mu} + \frac{\beta_2}{\alpha_2 - \mu} + \cdots + \frac{\beta_{N-1}}{\alpha_{N-1} - \mu}$$

and hence that

$$\langle h| f_N \rangle = \frac{-\omega_0}{\lambda\alpha_0 - 1} + \frac{\lambda\beta_1}{\lambda\alpha_1 - 1} + \frac{\lambda\beta_2}{\lambda\alpha_2 - 1} + \cdots + \frac{\lambda\beta_{N-1}}{\lambda\alpha_{N-1} - 1}.$$

Thus we see that the biorthogonal algorithm enables the Padé approximant of the integral equation to be represented as a continued fraction, and the sequence of $[N-1/N]$ Padé approximant solutions appear as the Nth convergent of an infinite continued fraction. The coefficients α_i and μ_i of the continued fraction are given initially by (3.16), and the role of the expansion parameter λ is preserved.

2.4 Potential Scattering

Many of the applications of Padé approximants have turned up in the theory of potential scattering, because potential-scattering theory provides a fine testing ground for the effectiveness of the Padé method on a profusion of interesting problems. It is also true that potential theory has proved a useful guide to the properties of quantum field theory. Indeed, because of the difficulty of calculating more than a few low-order terms of T-matrix elements in quantum field theory, Padé approximants may prove to be an essential tool for deriving reasonable extrapolations from a few terms of power series. We present here in this section the basic elements of scattering theory needed in our applications in quantum mechanics. We shall follow the notation and normalization of the book by Newton [1966, Chapter 7] where possible, because this book emphasizes K-matrix theory and Jost functions which are a necessary component of the theory of convergence of the Padé method; see also Graves-Morris [1973].

We use Schrödinger's equation in the two-particle relative coordinate $\mathbf{r}$, and take $\hbar = 1$. The reduced mass of the system, $m = (m_1^{-1} + m_2^{-1})^{-1}$, is taken to be $\frac{1}{2}$, defining the mass scale, so the equation takes the form

$$H\psi = -\nabla^2\psi(\mathbf{r}) + V(r)\psi(\mathbf{r}) = k^2\psi(\mathbf{r}). \qquad (4.1)$$

We assume that $V(r)$ is a short-range, spherically symmetric potential, which means that a positive μ exists for which $|e^{\mu r}V(r)|$ is bounded as $r \to \infty$. With these hypotheses, the equation (4.1) has scattering solutions of the type

$$\psi^{\text{asy}}(\mathbf{r}) \simeq e^{i\mathbf{k}\cdot\mathbf{r}} + f(\theta)\frac{e^{ikr}}{r} \qquad \text{as} \quad r \to \infty. \qquad (4.2)$$

Figure 1 shows the wave fronts in a pictorial representation of the scattering process, with an incident plane wave and a scattered wave.

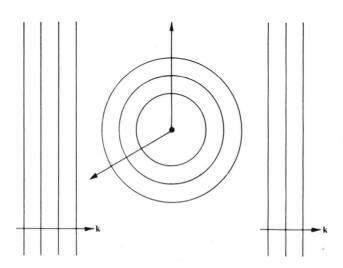

Figure 1. A pictorial representation of scattering, showing an incident wave of momentum $\mathbf{k}$ and an outgoing spherical wave.

One of the observable effects of the scattering process is the scattering cross-section, defined as the number of particles scattered per second per unit solid angle per unit incident flux, which is expressed in terms of the scattering amplitude $f(\theta)$ by the formula

$$\frac{d\sigma}{d\Omega} = |f(\theta)|^2.$$

To solve the Schrödinger equation, we write it as

$$(\nabla^2 + k^2)\psi(\mathbf{r}) = V(r)\psi(\mathbf{r})$$

and use the free Green's function, which is a solution of the equation

$$(\nabla^2 + k^2)G(\mathbf{r}, \mathbf{r}', k^2) = \delta(\mathbf{r} - \mathbf{r}') \tag{4.3}$$

representing propagation from the point $\mathbf{r} = \mathbf{r}'$. The solution we need is

$$G(\mathbf{r}, \mathbf{r}', k^2) = -\frac{1}{4\pi} \frac{e^{ik|\mathbf{r} - \mathbf{r}'|}}{|\mathbf{r} - \mathbf{r}'|}, \tag{4.4}$$

and the solution of Schrödinger's equation is given by the Born equation

$$\psi(\mathbf{r}) = e^{i\mathbf{k}\cdot\mathbf{r}} - \int \frac{e^{ik|\mathbf{r} - \mathbf{r}'|}}{4\pi|\mathbf{r} - \mathbf{r}'|} V(r')\psi(\mathbf{r}')\, d\mathbf{r}'. \tag{4.5}$$

We consider the form of $\psi(\mathbf{r})$ at a representative point $\mathbf{r}$ far from the scattering center. Let $\hat{\mathbf{r}}$ be a unit vector in the direction of $\mathbf{r}$, and $\mathbf{k}'$ be the momentum associated with scattering in this direction, so that $\mathbf{k}' = k\hat{\mathbf{r}}$. Then the asymptotic form of $\psi(\mathbf{r})$ follows from the Born equation, and is

$$\psi^{\text{asy}}(\mathbf{r}) = e^{i\mathbf{k}\cdot\mathbf{r}} - \frac{e^{ikr}}{4\pi r} \int e^{-i\mathbf{k}'\cdot\mathbf{r}'} V(r')\psi(\mathbf{r}')\, d\mathbf{r}' + O\left(\frac{1}{r^2}\right), \tag{4.6}$$

provided $V(r)$ is a short-range potential. This condition also implies that $|\psi(\mathbf{r})|$ is bounded as $r \to \infty$, and by comparing (4.2) and (4.6) we are led to define

$$T(\mathbf{k}', \mathbf{k}) = \langle \mathbf{k}'|T|\mathbf{k}\rangle = -4\pi f(\theta)$$

$$= \int e^{-i\mathbf{k}'\cdot\mathbf{r}} V(r)\psi(\mathbf{r})\, d\mathbf{r} \tag{4.7}$$

as the scattering amplitude, and the momentum-space representation of the potential

$$V(\mathbf{k}', \mathbf{k}) = \int e^{-i\mathbf{k}'\cdot\mathbf{r}} V(r)e^{i\mathbf{k}\cdot\mathbf{r}}\, d\mathbf{r}$$

$$= \langle \mathbf{k}'|V|\mathbf{k}\rangle. \tag{4.8}$$

We can solve for the free Green's function by momentum-space methods, taking care with the boundary conditions, and find as in (4.4) that

$$G(\mathbf{r},\mathbf{r}',k^2)=\frac{e^{ik|\mathbf{r}-\mathbf{r}'|}}{-4\pi|\mathbf{r}-\mathbf{r}'|}=\lim_{\varepsilon\downarrow 0}\int\frac{d\mathbf{q}}{(2\pi)^3}\frac{e^{i\mathbf{q}\cdot(\mathbf{r}-\mathbf{r}')}}{k^2-q^2+i\varepsilon}. \tag{4.9}$$

From (4.5), (4.7), and (4.9),

$$T(\mathbf{k}',\mathbf{k})=V(\mathbf{k}',\mathbf{k})+\frac{1}{(2\pi)^3}\int V(\mathbf{k}',\mathbf{q})\frac{d\mathbf{q}}{k^2-q^2+i\varepsilon}T(\mathbf{q},\mathbf{k}). \tag{4.10}$$

This equation is the Lippmann–Schwinger equation for the half-off-shell T-matrix, which is defined by (4.10) for all real $\mathbf{k}'$. We refer to Reed and Simon [1979, p. 98] for a discussion of how the Lippmann–Schwinger equation is incorporated in rigorous scattering theory. The on-shell T-matrix follows by taking $|\mathbf{k}'|=|\mathbf{k}|$. In three-body theory, it is useful to define a totally off-shell T-matrix by

$$T(\mathbf{k}',k^2,\mathbf{k}'')=V(\mathbf{k}',\mathbf{k}'')+\frac{1}{(2\pi)^3}\int V(\mathbf{k}',\mathbf{q})\frac{d\mathbf{q}}{k^2-q^2+i\varepsilon}T(\mathbf{q},k^2,\mathbf{k}''),$$

$$\tag{4.11}$$

and we shall return to this in Section 2.8. Equations (4.10) and (4.11) may be written schematically as

$$T=V+VGT=(1-VG)^{-1}V$$

$$=V+TGV=V(1-GV)^{-1}$$

$$=V+VGV+VGVGV+\cdots, \tag{4.12}$$

but these formulas do no more than denote algebraic manipulations on Equations (4.10) and (4.11). Provided the inverses exist and the series converge, the abbreviations are useful. For actual calculations, it is essential to reduce the dimension of (4.10) by expanding in partial waves, by means of the following projections:

$$t_{k',k}^{(l)}=-\frac{1}{8\pi}\int_{-1}^{1}T(\mathbf{k}',\mathbf{k})P_l(\cos\theta_{\mathbf{k}',\mathbf{k}})d(\cos\theta_{\mathbf{k}',\mathbf{k}}), \tag{4.13a}$$

$$v_{k',k}^{(l)}=-\frac{1}{8\pi}\int_{-1}^{1}V(\mathbf{k}',\mathbf{k})P_l(\cos\theta_{\mathbf{k}',\mathbf{k}})d(\cos\theta_{\mathbf{k}',\mathbf{k}}), \tag{4.13b}$$

so that

$$V(\mathbf{k}',\mathbf{q}) = -4\pi \sum_{l=0}^{\infty} (2l+1) P_l(\cos\gamma) v_{k',q}^{(l)} \qquad (4.14a)$$

and

$$T(\mathbf{k}',\mathbf{k}) = -4\pi \sum_{l=0}^{\infty} (2l+1) P_l(\cos\beta) t_{k',k}^{(l)}. \qquad (4.14b)$$

We have used $\gamma = \angle(\mathbf{k}',\mathbf{q})$ and $\beta = \angle(\mathbf{k}',\mathbf{k})$ in (4.14), and we need $\alpha = \angle(\mathbf{k},\mathbf{q})$ as well. These angles are shown in Figure 2. We need the "addition formula"

$$P_l(\cos\gamma) = P_l(\cos\alpha) P_l(\cos\beta)$$
$$+2\sum_{m=1}^{l} \frac{(l-m)!}{(l+m)!} P_l^m(\cos\alpha) P_l^m(\cos\beta) \cos m\phi$$

to give

$$t_{k'k}^{(l)} = v_{k'k}^{(l)} - \frac{2}{\pi} \int_0^{\infty} v_{k'q}^{(l)} \frac{q^2 dq}{k^2 - q^2 + i\varepsilon} t_{qk}^{(l)} \qquad (4.15)$$

This equation is the partial-wave Lippmann–Schwinger equation [Goldberger and Watson, 1964, p. 918], given formally by

$$T = V + VGT, \qquad (4.16)$$

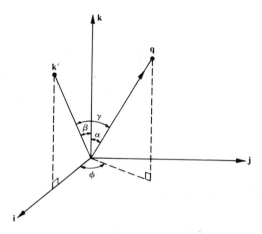

Figure 2. Angles between vectors $\mathbf{k}'$, $\mathbf{q}$, and $\mathbf{k}$, used to describe the scattering process.

just as in (4.12). To understand it better, let us consider the cases of exponential and Yukawa potentials. If V is an exponential potential,

$$V(r)=\lambda e^{-\mu r}.$$

With the momentum transfer Δ given by $\Delta^2 = (\mathbf{k}-\mathbf{k}')^2$, the potential is represented by

$$V(\mathbf{k}',\mathbf{k})=\frac{8\pi\lambda\mu}{\left(\Delta^2+\mu^2\right)^2}$$

in momentum space, and its S-wave component is given by

$$v_{k'k}^{(0)}=\frac{-2\lambda\mu}{\left[(k'+k)^2+\mu^2\right]\left[(k'-k)^2+\mu^2\right]}. \tag{4.17}$$

This takes on an especially simple form on the mass shell, and in general allows exact calculations with (4.15). If V is a Yukawa potential,

$$V(r)=\lambda e^{-\mu r}/r,$$

then

$$V(\mathbf{k}',\mathbf{k})=\frac{4\pi\lambda}{(\mathbf{k}'-\mathbf{k})^2+\mu^2}$$

and

$$v_{k'k}^{(l)}=\frac{-\lambda}{2kk'}Q_l\left(\frac{k'^2+k^2+\mu^2}{2kk'}\right) \tag{4.18}$$

gives the partial-wave projection in terms of the lth second-type Legendre function. These functions $v_{k'k}^{(l)}$ are well behaved ℓ^2 functions which cause no difficulties in (4.15). But the partial-wave Lippmann–Schwinger equation (4.15) is a singular integral equation (in the ℓ^2 Hilbert space), because the full kernel

$$S(k',q)=-\frac{2}{\pi}v_{k'q}^{(l)}\frac{q^2}{k^2-q^2+i\varepsilon}$$

does not satisfy the ℓ^2 condition

$$\int\int|S(k',q)|^2\,dk'\,dq<\infty,$$

and so Hilbert–Schmidt methods do not apply directly. However, the formal techniques of (4.12) may be valid for an iterative solution. In fact, the key requirement we need is that $t_{qk}^{(l)}$ be differentiable in q (or a Hölder continuity condition is sufficient) so that the right-hand side of (4.15) is well defined. Hence we are led to use a suitably restricted Banach space for $t_{qk}^{(l)}$. Then VG is a compact operator on the space, $(1-VG)^{-1}$ is a well-defined operator on the space, and the usual theory of integral equations with compact kernels is applicable.

Let us define the partial-wave K-matrix by the equation

$$K_{pq}=v_{pq}-\frac{2}{\pi}\int K_{pk'}\frac{k'^2\,dk'}{k^2-k'^2}v_{k'q}. \qquad (4.19)$$

The integral is a principal-value integral, and the superscript l for partial waves is suppressed. A Fourier transform shows that the Green's function analogous to (4.4) for the three-dimensional K-matrix equation is

$$G_r(\mathbf{r},\mathbf{r}',k^2)=\frac{\cos k|r-r'|}{-4\pi|r-r'|}=\frac{1}{(2\pi)^3}\int e^{i\mathbf{k}'\cdot(\mathbf{r}-\mathbf{r}')}\frac{d\mathbf{k}'}{k^2-k'^2}, \qquad (4.20)$$

which is real, as is denoted by the subscript r. The wave functions in each partial wave are real and are standing waves, and so the K-matrix is sometimes called the reaction matrix, or R-matrix for short. Equation (4.19) may be written formally as

$$K=V+KG_rV=(1+KG_r)V. \qquad (4.21)$$

Operating on each side of (4.16) with $(1+KG_r)$ on the left gives

$$(1+KG_r)T=K+KGT$$

and

$$T=K+K(G-G_r)T.$$

This equation is interpreted as

$$t_{pk}=K_{pk}-\frac{2}{\pi}\int K_{pq}\left(\frac{q^2}{k^2-q^2+i\varepsilon}-\frac{q^2}{k^2-q^2}\right)t_{qk}\,dq$$

$$=K_{pk}+ikK_{pk}t_{kk} \qquad (4.22)$$

and the on-shell partial-wave T-matrix is given by

$$t_{kk}=K_{kk}(1-ikK_{kk})^{-1}. \qquad (4.23)$$

Writing $K_{kk}^{(l)} = (\tan \delta_l)/k$, we find from (4.23) that

$$t_{kk}^{(l)} = \frac{e^{i\delta_l} \sin \delta_l}{k}. \tag{4.24}$$

Furthermore, (4.6), (4.7), and (4.14b) imply that

$$\psi^{asy}(\mathbf{r}) = e^{i\mathbf{k} \cdot \mathbf{r}} + \frac{1}{r} \sum_{l=0}^{\infty} (2l+1) e^{ikr + i\delta_l} \sin \delta_l \, P_l(\cos \theta), \tag{4.25}$$

and this equation shows why δ_l is called the phase shift of the lth partial wave.

We have approached the scattering problem by first finding the momentum-space representation and then going to partial waves. Frequently there are advantages in using partial waves *ab initio*, and then we will need the partial-wave Green's functions explicitly.

If the wave function is separated into partial waves,

$$\psi(\mathbf{r}) = \sum_{l=0}^{\infty} \sum_{m=-l}^{l} a_{lm} P_l^{|m|}(\cos \theta) e^{im\phi} R_l(r), \tag{4.26}$$

then the radial part $R_l(r)$ satisfies Schrödinger's equation in the form

$$\frac{1}{r} \frac{d^2}{dr^2} [rR_l(r)] - \frac{l(l+1)}{r^2} R_l(r) + k^2 R_l(r) = V(r) R_l(r), \tag{4.27}$$

and the substitution

$$\psi_l(r) = r R_l(r) \tag{4.28}$$

reduces (4.27) to

$$\frac{d^2 \psi_l(r)}{dr^2} - \frac{l(l+1)}{r^2} \psi_l(r) + k^2 \psi_l(r) = V(r) \psi_l(r). \tag{4.29}$$

If $V(r) = 0$, the non-interacting solutions of this equation obey

$$\frac{d^2 \psi_l(r)}{dr^2} - \frac{l(l+1)}{r^2} \psi_l(r) + k^2 \psi_l(r) = 0. \tag{4.30}$$

The independent solutions are $u_l(kr)$ and $v_l(kr)$, which are defined in terms of Bessel and Riccati–Bessel functions by

$$u_l(x) = \sqrt{\tfrac{1}{2}\pi x} \, J_{l+1/2}(x) = x j_l(x) \tag{4.31a}$$

and

$$v_l(x) = \sqrt{\tfrac{1}{2}\pi x}\, N_{l+1/2}(x) = x n_l(x). \tag{4.31b}$$

From (4.28) and (4.33a) we see that the radial part of an incident plane wave must be the solution of (4.29) vanishing at $r=0$, which is $u_l(kr)$. In fact, an incident plane wave has the expansion

$$e^{i\mathbf{k}\cdot\mathbf{r}} = \frac{1}{kr} \sum_{l=0}^{\infty} (2l+1) P_l(\cos\theta_{\mathbf{k},\mathbf{r}}) u_l(kr) i^l \tag{4.32}$$

and may be verified by using $(d/d\cos\theta)^l$ and then $r=0$. It is conventional to take account of the constant phase i^l explicitly in each partial wave, so that the incident wave is real.

We will need the following facts about $u_l(x)$ and $v_l(x)$, which are easily derived from well-known properties of Bessel functions:

$$u_l(x) \sim \frac{x^{l+1}}{(2l+1)!!} + O(x^{l+3}) \qquad \text{as} \quad x \to 0, \tag{4.33a}$$

$$v_l(x) \sim \frac{-x^{-l}}{(2l-1)!!} + O(x^{-l+2}) \qquad \text{as} \quad x \to 0, \tag{4.33b}$$

$$u_l(x) \sim \sin(x - \tfrac{1}{2}\pi l) \qquad \text{as} \quad x \to +\infty, \tag{4.34a}$$

$$v_l(x) \sim -\cos(x - \tfrac{1}{2}\pi l) \qquad \text{as} \quad x \to +\infty. \tag{4.34b}$$

The Wronskian of $u_l(x)$ and $v_l(x)$ is defined by

$$W(u_l, v_l) = \begin{vmatrix} u_l(x) & v_l(x) \\ u_l'(x) & v_l'(x) \end{vmatrix} = u_l(x) v_l'(x) - v_l(x) u_l'(x).$$

Because u_l and v_l satisfy the same second-order ordinary differential equation, it follows that $W(u_l, v_l)$ is independent of x, and either (4.33) or (4.34) shows that

$$W(u_l, v_l) = 1. \tag{4.35}$$

We need the further formulas and definitions

$$w_l^+(x) = -e^{i\pi l}[v_l(x) - i u_l(x)] \tag{4.36}$$

$$\sim e^{ix + \frac{1}{2}i\pi l} \qquad \text{as} \quad x \to +\infty. \tag{4.37}$$

We see that $w_l^+(x)$ represents an outgoing, scattered wave and is irregular at $x=0$. From (4.36),

$$W(u_l, w_l^+) = e^{i\pi(l+1)}. \qquad (4.38)$$

As a simple example of these functions, we have the S-wave formulas for (4.31) and (4.36):

$$u_0(x) = \sin x, \qquad v_0(x) = -\cos x, \quad \text{and} \quad w_0^+(x) = e^{ix}.$$

The solution of the partial-wave Schrödinger equation (4.29) is found by Green's functions to be

$$\psi_l(r) = u_l(kr) + \int_0^\infty g(r, r', k^2) V(r') \psi_l(r') \, dr', \qquad (4.39)$$

provided the Green's function is chosen so that

(i) $$\qquad \frac{\partial^2 g}{\partial r^2} - \frac{l(l+1)}{r^2} g + k^2 g = \delta(r-r'), \qquad (4.40)$$

(ii) $$\qquad g(r, r', k^2) \propto e^{ikr} \qquad \text{as} \quad r \to \infty,$$

(iii) $$\qquad g(0, r') = 0,$$

so that $\psi_l(r)$ is regular at the origin and the interaction leads to scattered waves only. We deduce that

$$g(r, r', k^2) = f_1(r') u_l(kr) \qquad \text{for} \quad r < r'$$
$$g(r, r', k^2) = f_2(r') w_l^+(kr) \qquad \text{for} \quad r > r' \qquad (4.41)$$

and to satisfy (4.40) at $r = r'$,

$$\left(\frac{\partial g}{\partial r} \right)_{r=r'+} - \left(\frac{\partial g}{\partial r} \right)_{r=r'-} = 1,$$

i.e.,

$$f_2(r) w_l^{+\prime}(kr) - f_1(r) u_l'(kr) = k^{-1}.$$

From (4.38) we deduce that

$$f_1(r) = -e^{-i\pi l} k^{-1} w_l^+(kr),$$
$$f_2(r) = -e^{i\pi l} k^{-1} u_l(kr),$$

which allows (4.41) to be written compactly as

$$g^{(l)}(r,r',k^2)=-e^{-i\pi l}u_l(kr_<)w_l^+(kr_>)/k.\tag{4.42}$$

As an example, the standard Green's function for scattering in S-waves is

$$g^{(0)}(r,r',k^2)=-k^{-1}\sin kr_< \, e^{ikr_>}.$$

We may deduce from (4.42) and (4.36) that K-matrix (standing-wave) Green's function, which is real, is

$$g_r^{(l)}(r,r',k^2)=u_l(kr_<)v_l(kr_>)/k.\tag{4.43}$$

As an example, the standing-wave Green's function for the reaction in S-waves is

$$g_r^{(0)}(r,r',k^2)=-k^{-1}\sin kr_< \cos kr_>.$$

We substitute (4.42) in (4.39) to obtain the asymptotic form

$$\psi_l^{\mathrm{asy}}(r)=\sin(kr-\tfrac{1}{2}\pi l)-k^{-1}e^{ikr-\frac{1}{2}i\pi l}\int_0^\infty u_l(kr')V(r')\psi_l(r')\,dr'.\tag{4.44}$$

Just as with (4.6), we are led to define

$$\tau_l=-k^{-1}\int_0^\infty u_l(kr')V(r')\psi_l(r')\,dr',\tag{4.45}$$

and then

$$\psi_l^{\mathrm{asy}}(r)=\frac{e^{ikr-\frac{1}{2}i\pi l}(1+2i\tau_l)-e^{-ikr+\frac{1}{2}i\pi l}}{2i},\tag{4.46}$$

in which the first and second terms represent the outgoing wave and incoming wave respectively. Conservation of flux between these waves requires that

$$|1+2i\tau_l|=1,$$

and it is usual to define

$$S_l=1+2i\tau_l=e^{2i\delta_l}\quad\text{and}\quad\tau_l=e^{i\delta_l}\sin\delta_l.\tag{4.47}$$

Of course, the phase shifts of (4.47) are precisely the same as those of (4.25) by virtue of (4.26), (4.28), (4.32), and (4.34a), which amount to (4.46).

To complete our discussion of partial-wave Green's functions, we need the Green's function for Jost solutions of the partial-wave Schrödinger equation (4.29). These are the solutions

$$f(k, r) \sim e^{ikr} \quad \text{and} \quad f(-k, r) \sim e^{-ikr} \tag{4.48}$$

representing outgoing and incoming waves at infinity. Each is an unphysical solution because it does not conserve flux. The solutions are given by

$$f_l(k, r) = e^{ikr} + \int_0^\infty \tilde{g}_l(r, r', k^2) V(r') f_l(k, r') \tag{4.49a}$$

and

$$f_l(-k, r) = e^{-ikr} + \int_0^\infty \tilde{g}_l(r, r', k^2) V(r') f_l(-k, r'). \tag{4.49b}$$

We need the Green's function satisfying

$$\tilde{g}(r, r', k^2) = 0 \qquad \text{for} \quad r > r', \tag{4.50a}$$

and a derivation similar to that of (4.42) but using (4.35) instead of (4.38) gives

$$\tilde{g}_l(r, r', k^2) = \frac{1}{k} \{ v_l(kr') u_l(kr) - u_l(kr') v_l(kr) \} \qquad \text{for} \quad r < r'. \tag{4.50b}$$

For the particular case of S-waves needed in (9.3) below,

$$\tilde{g}(r, r', k^2) = -k^{-1} \sin k(r - r') \qquad \text{for} \quad r < r',$$
$$\tilde{g}(r, r', k^2) = 0 \qquad \text{for} \quad r > r',$$

giving rise to the Volterra integral equations for the S-wave:

$$f(k, r) = e^{ikr} - k^{-1} \int_r^\infty \sin k(r - r') V(r') f(k, r') \, dr'$$

and   (4.51)

$$f(-k, r) = e^{-ikr} - k^{-1} \int_r^\infty \sin k(r - r') V(r') f(-k, r') \, dr'.$$

From these two independent solutions of Schrödinger's equation we may form the physical solution

$$\psi_l(r) = af(k, r) + bf(-k, r).$$

As $r \to \infty$, (4.46)–(4.48) show that $S_l = -(a/b)e^{i\pi l}$. At $r = 0$, $\psi_l(0) = 0$, and hence

$$S_l = \frac{f(-k,0)}{f(k,0)} e^{i\pi l}, \tag{4.52}$$

showing how the Jost functions determine the S-matrix.

Considerable attention has been paid to the solution of the Lippmann–Schwinger equation, (4.15), because it is a prototype equation for quantum scattering theory. The natural method of solution of the Lippmann–Schwinger equation using Padé approximants is to use a sequence of nearly diagonal approximants of the Liouville–Neumann series, as described in Section 2.2. The first numerical results for the Yukawa potential using this approach were given by Caser et al. [1969] and were most encouraging. We emphasize again (cf. Part I, Section 2.4, and Schwartz [1966]) that considerable numerical accuracy is required for the coefficients of the Liouville–Neumann series when Padé approximation is used. In this chapter, we discuss various developments of this direct approach to the Lippmann–Schwinger equation. It is known [Lovelace, 1964; Graves-Morris and Rennison, 1974] that the kernel of the Lippmann–Schwinger equation for Yukawa scattering is compact on a suitably chosen Banach space. In the light of the discussion of Section 2.2, it is therefore not surprising that the direct Padé method is successful in this application.

A variety of quite distinct Padé methods have been used to solve the Lippmann–Schwinger equation and related equations. In Section 2.7, we prove that the direct Padé method converges for single-sign ordinary potentials. In Section 2.8, we discuss variational Padé methods which normally improve the accuracy of the direct Padé method. In Section 2.9, we discuss singular potentials. An interesting approach to low-energy scattering problems is to start with the doubly off-shell partial-wave K-matrix Lippmann–Schwinger equation (4.19),

$$K_{pq}(k^2) = v_{pq} + \frac{2}{\pi} \int_0^\infty K_{pk'}(k^2) \frac{k'^2\, dk'}{k'^2 - k^2} v_{k'q}.$$

The kernel, as a Hilbert-space kernel, is nonsingular for $k^2 \le 0$, and the equation is easily solved by matrix inversion. The direct Padé method necessarily converges (see Section 2.7). The solution is evaluated for fixed p, q, and N different negative values of k^2. The solution is then extrapolated, using N-point rational interpolants, to the scattering region, $k^2 > 0$ [Schlessinger and Schwartz, 1966]. A similar extrapolation is also useful for the Bethe–Salpeter equation [Haymaker, 1968; Haymaker and Schlessinger, 1970].

The Faddeev equations describe quantum-mechanical three-body potential scattering. They are integral equations, involving double integrals over a kernel containing polar and logarithmic singularities, and are usually coupled integral equations in cases of physical interest. The Faddeev equations are singular in such a complicated way that solution by formation of Padé approximants to the Liouville–Neumann series is a very attractive method [Tjon 1970, 1973, 1977].

For further details of the role of Padé approximants in particular approaches to the Lippmann–Schwinger equation, we refer to Ruijgrok [1968], Stern and Warburton [1972], and Warburton and Stern [1968].

2.5 Derivation of Padé Approximants from Variational Principles

Variational principles have often turned out to lead to the most accurate methods of solving the Schrödinger equation in bound-state and scattering problems. Certainly it is true that quite drastic approximations to the wave functions lead to surprisingly accurate values of the binding energy in the application of the Rayleigh–Ritz method.

We first discuss bound-state problems in quantum mechanics, which are directly associated with homogeneous integral equations. These are to be distinguished carefully from scattering problems, associated with inhomogeneous integral equations, which we treat second and separately.

Bound-state problems usually involve the determination of discrete energy levels. The associated integral equations only have solutions for particular discrete values of a parameter occurring in the theory. Although it is not equivalent to Padé approximation, we consider first the Rayleigh–Ritz variational method. We will then use this method with a particular approach which yields Padé approximants as the variational solution. We consider a potential $V(r)$ which is at least partly attractive. The Rayleigh–Ritz principle shows that such a potential has at least one bound state, and we wish to determine the energy of the lowest-lying bound state.

Schrödinger's equation for the wave function ψ of a particle in a potential V may be written as

$$H\psi \equiv -\nabla^2\psi + V\psi \equiv H_0\psi + V\psi = E\psi \tag{5.1}$$

following (4.1). The Rayleigh–Ritz principle is the statement that

$$[E] = \frac{\langle \psi_t | H | \psi_t \rangle}{\langle \psi_t | \psi_t \rangle}, \tag{5.2}$$

which takes real values for any normalizable function $|\psi_t\rangle$, has a minimum value with respect to variations of the trial function $|\psi_t\rangle$. This minimum value of $[E]$ is the ground-state binding energy, and the associated trial

function is the bound-state wave function. The proof of the principle is easy
if the existence of a complete orthonormal set of eigenfunctions of H is
assumed. The variational method consists of selecting M functions

$$\psi_i(\mathbf{x}), \qquad i=1,2,\ldots, M, \tag{5.3}$$

and forming from these a trial wave function

$$\psi_t(\mathbf{x})= \sum_{i=1}^{M} \alpha_i\psi_i(\mathbf{x}). \tag{5.4}$$

As an example, we might choose

$$\psi_t(\mathbf{x})=\alpha_1 e^{-|x|} +\alpha_2 e^{-3|x|}.$$

Consequently, $[E]$ becomes a function of $\{\alpha_i\}$; the values of $\{\alpha_i\}$ which give
a minimum value of $[E]$ determine the best approximation to the true wave
function, and the actual minimum value of $[E]$ is the estimated binding
energy. Another form of the variational method allows each $\psi_i(\mathbf{x})$ to depend
on other parameters

$$\{\beta_{ij}\}, \qquad j=1,2,\ldots, M_i,$$

which may also be varied so as to give a minimum of $[E]$. For example

$$|\psi_t\rangle =\alpha_1 |x|^{\beta_{11}} e^{-\beta_{12}|x|} +\alpha_2 |x|^{\beta_{21}} e^{-3|x|}$$

may be varied with respect to the five variational parameters to yield a
minimum value of $[E]$ and an approximation to $\psi(\mathbf{x})$.

The Rayleigh–Ritz method has been used for very precise calculations of
the binding energy of atomic helium, where the significant forces are
electromagnetic and are accurately known [Pekeris 1958, 1959]. The method
has also been used for the calculation of the binding energy of the triton
using nuclear potentials, where the results are more of a test of reliability of
the potentials [Delves and Phillips, 1969].

Having considered the basic Rayleigh–Ritz principle, we contrast this
procedure with another variational principle, and see in what form this
approach yields Padé approximants as the variational solution. The bound-
state problem is reformulated to be the determination of the potential
strength that gives a particular bound-state energy. The potential strength is
measured by a parameter λ, and we take

$$V(r)=\lambda U(r) \tag{5.5}$$

and consider

$$\left[-\frac{1}{\lambda}\right] = \frac{\langle\psi_t|U(H_0-E)^{-1}U|\psi_t\rangle}{\langle\psi_t|U|\psi_t\rangle}. \tag{5.6}$$

First we notice that if $|\psi_t\rangle$ is a solution of the Schrödinger equation (5.1), then $[-1/\lambda] = -1/\lambda$. Let us define $G = (H_0-E)^{-1}$ as the usual free Green's function and consider first-order variations in $[-1/\lambda]$,

$$\begin{aligned}
\delta\left[\frac{-1}{\lambda}\right] &= \frac{\langle\psi_t|UGU|\delta\psi_t\rangle}{\langle\psi_t|U|\psi_t\rangle} + \frac{\langle\delta\psi_t|UGU|\psi_t\rangle}{\langle\psi_t|U|\psi_t\rangle} \\
&\quad - \langle\psi_t|UGU|\psi_t\rangle\langle\delta\psi_t|U|\psi_t\rangle[\langle\psi_t|U|\psi_t\rangle]^{-2} \\
&\quad - \langle\psi_t|UGU|\psi_t\rangle\langle\psi_t|U|\delta\psi_t\rangle[\langle\psi_t|U|\psi_t\rangle]^{-2}. \tag{5.7}
\end{aligned}$$

Since the exact solution $|\psi\rangle$ satisfies

$$GV|\psi\rangle = -|\psi\rangle, \tag{5.8}$$

it follows that

$$\delta\left[\frac{-1}{\lambda}\right] = 0 \qquad \text{to first order.}$$

Hence we say that $[-1/\lambda]$ is stationary at

$$|\psi_t\rangle = |\psi\rangle. \tag{5.9}$$

Quite clearly, a great deal more needs to be done to give conditions to make the previous formulas rigorous. But provided $\langle\psi_t|U|\psi_t\rangle$ is sufficiently close to $\langle\psi|U|\psi\rangle$, the rigorous discussion adds little to the understanding. In conclusion, this variational method consists of finding a turning point of $[-1/\lambda]$ by varying the parameters of $|\psi_t\rangle$, and it gives the value of the coupling strength λ which produces a bound state of given energy.

We use the following set of trial functions [Nuttall, 1966, 1967, 1970a]:

$$|\psi_t\rangle = \sum_{i=0}^{M-1} d_i(GU)^i|\psi_0\rangle, \tag{5.10}$$

and the coefficients $d_0, \ldots, d_{N-1}$ are to be determined. Then

$$\left[\frac{-1}{\lambda}\right] = \frac{\displaystyle\sum_{i=0}^{M-1}\sum_{j=0}^{M-1} d_i d_j^* \langle\psi_0|U(GU)^{i+j+1}|\psi_0\rangle}{\displaystyle\sum_{i=0}^{M-1}\sum_{j=0}^{M-1} d_i d_j^* \langle\psi_0|U(GU)^{i+j}|\psi_0\rangle}. \tag{5.11}$$

We make $[-1/\lambda]$ stationary by requiring

$$\frac{\partial}{\partial d_i}\left[\frac{-1}{\lambda}\right]=0 \qquad \text{for} \quad i=0,1,\dots M-1 \qquad (5.12)$$

and define

$$c_k=\langle\psi_0|U(GU)^k|\psi_0\rangle, \qquad k=0,1,2,\dots. \qquad (5.13)$$

Substituting from (5.11) into (5.12), we find the set of linear equations

$$\sum_{j=0}^{M-1} d_j^* c_{i+j+1}-\left[\frac{-1}{\lambda}\right]\sum_{j=0}^{M-1} d_j^* c_{i+j}=0.$$

The elements c_k of (5.13) are real in the bound-state problem, and hence we see that $d_0,\dots,d_{M-1}$ are determined by

$$\begin{pmatrix} c_0+[\lambda]c_1 & c_1+[\lambda]c_2 & \cdots & c_{M-1}+[\lambda]c_M \\ c_1+[\lambda]c_2 & c_2+[\lambda]c_3 & \cdots & c_M+[\lambda]c_{M+1} \\ \vdots & \vdots & & \vdots \\ c_{M-1}+[\lambda]c_M & c_M+[\lambda]c_{M+1} & \cdots & c_{2M-2}+[\lambda]c_{2M-1} \end{pmatrix}\begin{pmatrix} d_0 \\ d_1 \\ \vdots \\ d_{M-1} \end{pmatrix}^*=0.$$

$$(5.14)$$

These equations determine $d_0:d_1:\cdots:d_{N-1}$ and $[\lambda]\equiv[1/\lambda]_{st}^{-1}$ from the consistency condition that

$$\begin{vmatrix} c_0+[\lambda]c_1 & c_1+[\lambda]c_2 & \cdots & c_{M-1}+[\lambda]c_M \\ c_1+[\lambda]c_2 & c_2+[\lambda]c_3 & \cdots & c_M+[\lambda]c_{M+1} \\ \vdots & \vdots & & \vdots \\ c_{M-1}+[\lambda]c_M & c_M+[\lambda]c_{M+1} & \cdots & c_{2M-2}+[\lambda]c_{2M-1} \end{vmatrix}=0.$$

$$(5.15)$$

We compare this method with the Padé method of solving the Schrödinger equation (5.1), which we write as a homogeneous integral equation

$$|\psi\rangle=-GV|\psi\rangle. \qquad (5.16)$$

To solve (5.16) we introduce a function $|\psi_0\rangle$ and parameters λ and η. λ retains its interpretation through (5.5), and we seek a solution of

$$|\psi\rangle=\eta|\psi_0\rangle-\lambda GU|\psi\rangle \qquad (5.17)$$

with $\eta=0$. Provided $\eta\neq0$, we find formally

$$|\psi\rangle=\eta\{|\psi_0\rangle-\lambda GU|\psi_0\rangle+(-\lambda GU)^2|\psi_0\rangle+\cdots\},$$

and

$$\langle\psi_0|U|\psi\rangle=\eta\{c_0-\lambda c_1+(-\lambda)^2c_2+\cdots\}.$$

Forming Padé approximants to the series $c_0+\lambda'c_1+\lambda'^2c_2+\cdots$, we have

$$Q^{[M-1/M]}(-\lambda)\langle\psi_0|U|\psi\rangle=\eta P^{[M-1/M]}(-\lambda)+O(\lambda^{2M}).$$

We see that a zero of $Q^{[M-1/M]}(-\lambda)$ is consistent with $\eta=0$ within the approximation scheme. Hence the Padé method boils down to the condition that $Q^{[M-1/M]}(-\lambda)=0$, which may be written [see (I.1.3.2)] as

$$\begin{vmatrix} c_0+\lambda c_1 & c_1+\lambda c_2 & \cdots & c_{M-1}+\lambda c_M \\ c_1+\lambda c_2 & c_2+\lambda c_3 & \cdots & c_M+\lambda c_{M+1} \\ \vdots & \vdots & & \vdots \\ c_{M-1}+\lambda c_M & c_M+\lambda c_{M+1} & \cdots & c_{2M-2}+\lambda c_{2M-1} \end{vmatrix}=0, \quad (5.18)$$

agreeing with (5.15). We conclude that the Padé method and the specific variational method of (5.10) and (5.11) give identical results.

Let us proceed to the problems of scattering theory, where the Kohn method for Green's function and the Schwinger method for the T-matrix are widely used. Suppose we wish to calculate matrix elements of the full interacting Green's function

$$I=\langle h|(E-H)^{-1}|g\rangle. \quad (5.19)$$

Then a stationary expression for I is

$$[I]=\langle h|\psi_t\rangle+\langle\psi_t'|g\rangle-\langle\psi_t'|E-H|\psi_t\rangle. \quad (5.20)$$

$[I]$ is a bivariational functional, which must be made stationary with respect to independent variations in $|\psi_t\rangle$ and $\langle\psi_t'|$. This leads to

$$\delta[I]=\langle h|\delta\psi_t\rangle-\langle\psi_t'|E-H|\delta\psi_t\rangle+\langle\delta\psi_t'|g\rangle$$

$$-\langle\delta\psi_t'|E-H|\psi_t\rangle-\langle\delta\psi_t'|E-H|\delta\psi_t\rangle. \quad (5.21)$$

Hence we see that $[I]$ is stationary if

$$|\psi_t\rangle = (E-H)^{-1}|g\rangle \tag{5.22a}$$

and

$$\langle \psi_t'| = \langle h|(E-H)^{-1}, \tag{5.22b}$$

and at the stationary point

$$[I] = \langle h|(E-H)^{-1}|g\rangle,$$

agreeing with (5.19). If we make the choices [Nuttall, 1966, 1967, 1970a; Conn, 1974]

$$|\psi_t\rangle = \sum_{i=0}^{M-1} d_i (GU)^i G|g\rangle \tag{5.23a}$$

and

$$\langle \psi_t'| = \sum_{i=0}^{M-1} d_i' \langle h|(GU)^i G, \tag{5.23b}$$

we can then find the stationary points of $[I]$:

$$[I] = \sum_{i=0}^{M-1} (d_i + d_i')\langle h|(GU)^i G|g\rangle$$
$$- \sum_{i=0}^{M-1}\sum_{j=0}^{M-1} d_i' d_j \langle h|(GU)^i[1-\lambda GU](GU)^j G|g\rangle,$$

and defining $c_i = \langle h|(GU)^i G|g\rangle$ for $i=0,1,\ldots$, this expression becomes

$$[I] = \sum_{i=0}^{M-1} (d_i + d_i')c_i - \sum_{i=0}^{M-1}\sum_{j=0}^{M-1} d_i' d_j (c_{i+j} - \lambda c_{i+j+1}),$$

$$\frac{\partial [I]}{\partial d_i} = 0 \quad \text{if} \quad c_i = \sum_{j=0}^{M-1} d_j' (c_{i+j} - \lambda c_{i+j+1}),$$

$$\frac{\partial [I]}{\partial d_i'} = 0 \quad \text{if} \quad c_i = \sum_{j=0}^{M-1} d_j (c_{i+j} - \lambda c_{i+j+1}).$$

These are $2M$ linear equations for $2M$ unknowns, $d_j, d_j', j=0,1,\ldots M-1$.

We define (as in (3.7)) the matrix

$$C_{ij} = c_{i+j} - \lambda c_{i+j+1},$$

and then (5.20) reduces to

$$[I] = \sum_{i=0}^{M-1} \sum_{j=0}^{M-1} c_i (C^{-1})_{ij} c_j. \tag{5.24}$$

This equation is the variational solution for the interacting Green's-function matrix element using the variations of (5.23). It is also the Padé-approximant solution obtained from expansion of the full Green's function in terms of the free Green's function

$$I = \langle h | G + GVG + GVGVG + \cdots | g \rangle.$$

As usual, we form Padé approximants to the series

$$I = \langle h | G + \lambda GUG + \lambda^2 GUGUG + \cdots | g \rangle$$
$$= c_0 + \lambda c_1 + \lambda^2 c_2 + \cdots,$$

and (I.1.3.6) shows that

$$[M-1/M]_I = \sum_{i=0}^{M-1} \sum_{j=0}^{M-1} c_i (C^{-1})_{ij} c_j,$$

confirming (5.24).

It is interesting to perform this calculation with any potential for which the full Green's function is known analytically. For the exponential potential,

$$V(r) = \lambda \exp(-r/a),$$

and the S-wave radial equation may be solved by making the substitution $x = \exp(-r/a)$. The full S-wave Green's function is [Basdevant and Lee, 1969b]

$$g(r, r', k^2) = \frac{-a\pi}{\sin \pi \nu} \frac{J_{-\nu}(\eta \zeta_>)}{J_{-\nu}(\eta)} [J_\nu(\eta) J_{-\nu}(\eta \zeta_<) - J_{-\nu}(\eta) J_\nu(\eta \zeta_<)]$$

where

$$\zeta_< = \exp(-r_</2a), \qquad \zeta_> = \exp(-r_>/2a),$$
$$\eta = 2a\sqrt{-\lambda}, \quad \text{and} \quad \nu = 2aik.$$

The relevant Bessel functions may be expanded in powers of η, leading to

$$g(r, r', k^2) = \sum_{i=0}^{\infty} \lambda^i c_i(r, r', k^2), \qquad (5.25)$$

where the coefficients c_i are known functions of r, r', and k^2. The poles of g are then given approximately by the zeros of $Q^{[M-1/M]}(\lambda)$. These zeros should be independent of r, r' from the mathematical theory of integral equations, or because physically they represent couplings for bound states. Accordingly, one makes a sensible choice, such as $r = r' = a$, upon which we will improve in the next section. The condition

$$Q^{[M-1/M]}(\lambda) = 0 \qquad (5.26)$$

then becomes an equation in the variables k, λ giving either the couplings for a bound state of given binding energy or, because we have an analytic solution, the binding energy for a given coupling.

The Schwinger variational method calculates the amplitude T given by

$$T = \langle h | V + V(E-H)^{-1}V | g \rangle \qquad (5.27)$$

from the bivariational functional

$$[T] = \langle \psi_t' | V | g \rangle + \langle h | V | \psi_t \rangle - \langle \psi_t' | V - VGV | \psi_t \rangle. \qquad (5.28)$$

Then we may prove, as with the Kohn principle, that $[T]$ is stationary with respect to all possible variations of $|\psi_t\rangle$ and $\langle \psi_t' |$ if $|\psi_t\rangle$ and $\langle \psi_t' |$ satisfy

$$\langle \psi_t' | = \langle h | + \langle \psi_t' | VG$$

and

$$|\psi_t\rangle = |g\rangle + GV|\psi_t\rangle. \qquad (5.29)$$

In this case,

$$[T] = \langle h | (1 - VG)^{-1} V | g \rangle = \langle h | V + V(E-H)^{-1}V | g \rangle$$
$$= T,$$

proving that $[T]$ is a bivariational functional for the T-matrix. For the scattering problem, we have the formal expansion for the T-matrix elements

$$T = \langle h | V + VGV + VGVGV + \cdots | g \rangle,$$

where $\langle h|$, $|g\rangle$ are usually plane-wave states or their partial-wave components. We may write

$$T = \lambda\langle h|U|g\rangle + \lambda^2\langle h|UGU|g\rangle + \cdots$$
$$= \lambda c_0 + \lambda^2 c_1 + \cdots$$

and form Padé approximants to T/λ. The result for the $[M-1/M]$ Padé approximant is precisely the same as for the Schwinger variational method using, [Garibotti, 1972],

$$|\psi_t\rangle = \sum_{i=0}^{M-1} d_i(GV)^i|g\rangle$$

and

$$\langle\psi_t'| = \sum_{i=0}^{M-1} d_i\langle h|(VG)^i.$$

The analysis is very similar to that for the Kohn method and not worth repeating. We summarize by noting that for three particular variational methods, the solution obtained is the same as that of the $[M-1/M]$ Padé approximant.

To complete the discussion, let us observe that we can derive all Padé approximants of type $[N+J-1/N-J]$, for $J=0,1,2,\ldots,N$, which lie above the diagonal by restricting the choice of trial vectors. For example, in (5.22), the choices

$$|\psi_t\rangle = \sum_{i=0}^{J-1} \lambda^i(GU)^iG|g\rangle + \sum_{i=J}^{N-1} d_i(GU)^iG|g\rangle$$

and

$$\langle\psi_t'| = \sum_{i=0}^{J-1} \lambda^i\langle h|(GU)^iG + \sum_{i=J}^{N-1} d_i'\langle h|(GU)^iG$$

allow a similar analysis, leading to $2(N-J)$ equations for $2(N-J)$ unknowns $\{d_i, d_i'\}$ and eventually to the $[N+J-1/N-J]$ Padé approximant.

Reverting to the Rayleigh–Ritz principle at the beginning of the section, we consider a refinement which is a variational principle of Bessis [1976]. The Rayleigh–Ritz method and principle are summarized by the equations

$$[E] = \frac{\langle\psi_t|H|\psi_t\rangle}{\langle\psi_t|\psi_t\rangle},$$
$$E_0 = \min_{|\psi_t\rangle}[E].$$

If it so happens that some of the higher moments of the trial wave function $|\psi_t\rangle$ are available, namely

$$\mu_i = \langle \psi_t | H^i | \psi_t \rangle, \tag{5.30}$$

Bessis's method [1976] makes use of this information. We use a supertrial wave function

$$|\psi_s\rangle = \sum_{i=0}^{M-1} d_i H^i |\psi_t\rangle, \tag{5.31}$$

where $|\psi_t\rangle$ is the ordinary trial wave function and $\{d_i\}$ are variational parameters. The principle is now rewritten as

$$E_0 = \min_{|\psi_t\rangle} \min_{\{d_i\}} \frac{\langle \psi_s | H | \psi_s \rangle}{\langle \psi_s | \psi_s \rangle}, \tag{5.32}$$

indicating that the minimization with respect to $\{d_i\}$ is to be done first. Of course, the lowest bound-state energy E_0 would be given exactly if ever minimization over all trial wave functions $|\psi_t\rangle$ could be realized. But because in practice this is approximate, the minimization over $\{d_i\}$ may be worthwhile. Retaining the definition of $\{\mu_i\}$ given by (5.30), we find

$$[E] = \frac{\displaystyle\sum_{i=0}^{M-1} \sum_{j=0}^{M-1} d_i d_j^* \langle \psi_t | H^{i+j+1} | \psi_t \rangle}{\displaystyle\sum_{i=0}^{M-1} \sum_{j=0}^{M-1} d_i d_j^* \langle \psi_t | H^{i+j} | \psi_t \rangle}$$

$$= \frac{\displaystyle\sum_{i=0}^{M-1} \sum_{j=0}^{M-1} d_i d_j^* \mu_{i+j+1}}{\displaystyle\sum_{i=0}^{M-1} \sum_{j=0}^{M-1} d_i d_j^* \mu_{i+j}}. \tag{5.33}$$

We make $[E]$ stationary by requiring

$$\frac{\partial [E]}{\partial d_i} = 0 \qquad \text{for} \quad i = 0, 1, \ldots, M-1. \tag{5.34}$$

Substituting (5.33) into (5.34), we find the set of linear equations

$$\sum_{j=0}^{M-1} d_j^* \mu_{i+j+1} - [E] \sum_{j=0}^{M-1} d_j^* \mu_{i+j} = 0. \tag{5.35}$$

The moments $\{\mu_i\}$ are all real, and hence $[E]$ is determined by the consistency condition for the M homogeneous equations (5.35) for $d_0:d_1:d_2:\cdots:d_{M-1}$. This condition is

$$\begin{vmatrix} \mu_1-[E]\mu_0 & \mu_2-[E]\mu_1 & \cdots & \mu_M-[E]\mu_{M-1} \\ \mu_2-[E]\mu_1 & \mu_3-[E]\mu_2 & \cdots & \mu_{M+1}-[E]\mu_M \\ \vdots & \vdots & & \vdots \\ \mu_M-[E]\mu_{M-1} & \mu_{M+1}-[E]\mu_M & \cdots & \mu_{2M-1}-[E]\mu_{2M-2} \end{vmatrix}=0,$$

or alternatively

$$\begin{vmatrix} \mu_0 & \mu_1 & \cdots & \mu_M \\ \mu_1 & \mu_2 & \cdots & \mu_{M+1} \\ \vdots & \vdots & & \vdots \\ \mu_{M-1} & \mu_M & \cdots & \mu_{2M-1} \\ 1 & [E] & \cdots & [E]^M \end{vmatrix}=0. \tag{5.36}$$

Bessis's principle is that a minimum value of the solution $[E]$ of (5.36) is to be found over a set of trial wave functions $|\psi_t\rangle$ which determine $\{\mu_i\}$ through (5.30).

This is precisely the scheme which would emerge by the simplest Padé method. To find the poles in the E-plane of the Green's-function matrix element

$$\langle\psi_t|(E-H)^{-1}|\psi_t\rangle,$$

it is natural to use $w=E^{-1}$ and to write

$$\langle\psi_t|(E-H)^{-1}|\psi_t\rangle = w\langle\psi_t|1+wH+w^2H^2+\cdots|\psi_t\rangle$$

$$= \sum_{i=0}^{\infty} w^{i+1}\mu_i. \tag{5.37}$$

We see that the solution (5.36) of the variational problem is the position of the pole on the left-hand real w-axis nearest the origin given by the $[M/M]$ Padé approximant to $\langle\psi_t|(E-H)^{-1}|\psi_t\rangle$ of (5.37). The spectrum of the exact full Green's function is shown in fig. 1.

Probably the most useful solution is that with $m=2$, with a quadratic equation for $[E]$,

$$[E]^2\begin{vmatrix} \mu_0 & \mu_1 \\ \mu_1 & \mu_2 \end{vmatrix}+[E]\begin{vmatrix} \mu_2 & \mu_0 \\ \mu_3 & \mu_1 \end{vmatrix}+\begin{vmatrix} \mu_1 & \mu_2 \\ \mu_2 & \mu_3 \end{vmatrix}=0,$$

w-plane

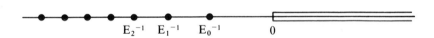

$$E_2^{-1} \quad E_1^{-1} \quad E_0^{-1} \qquad 0$$

Figure 1. The spectrum of the full Green's function.

which has an explicit solution in terms of $\mu_0, \mu_1, \mu_2, \mu_3$. Even so, Bessis's principle requires the minimization over the parameters of $|\psi_t\rangle$ of a nonlinear and complicated function. However, the variational justification of it shows that it must be at least as accurate as the simple Rayleigh–Ritz principle.

In this section we have concentrated on the connection between Padé approximants and variational principles rather than the bounding properties which may be derived under suitable conditions. In this context, we refer to Nuttall [1966, 1967, 1970a, 1977], Bessis and Villani [1975], Giraud [1978], and Giraud and Turchetti [1978]. For further details of the connection with Brillouin–Wigner and Rayleigh–Schrödinger perturbation theories, we refer to the review of Killingbeck [1977], and to Amos [1978], Benofy and Gammel [1977], Young et al. [1957], Schofield [1972], and Goldhammer and Feenberg [1956].

2.6 An Error Bound on Padé Approximants from Variational Principles

Surprisingly, variational methods such as the Kohn method, for which the error takes a simple form, can lead to (*a posteriori*) error estimates for the Padé solution to integral equations. We revert to the formalism of Section 2.3 for integral equations of the type

$$|f\rangle = |g\rangle + \lambda A |f\rangle \tag{6.1}$$

and the Hermitian-conjugate equation

$$\langle f'| = \langle h| + \lambda \langle f'|A. \tag{6.2}$$

To solve (6.1) for the quantity $\langle h|f\rangle$, we consider the variational quantity

$$[J] = \langle h|\psi_t\rangle + \langle \psi_t'|g\rangle - \langle \psi_t'|1 - \lambda A|\psi_t\rangle. \tag{6.3}$$

The stationary value of $[J]$ occurs when

$$|\psi_t\rangle = |f\rangle \quad \text{and} \quad \langle \psi_t'| = \langle f'|, \tag{6.4}$$

under which circumstances we find, as in Section 2.5, that

$$[J]_{st} = J = \langle h | f \rangle = \langle f' | g \rangle. \tag{6.5}$$

When $| \psi_t \rangle$ and $\langle \psi_t' |$ are not set according to (6.4), the variation of $[J]$ is given by

$$\delta[J] = [J] - J. \tag{6.6}$$

Recall that if

$$| \psi_t \rangle = \sum_{i=0}^{N-1} d_i A^i | g \rangle \tag{6.7a}$$

and

$$\langle \psi_t' | = \sum_{i=0}^{N-1} d_i' \langle h | A^i \tag{6.7b}$$

are the variational solutions of (6.3), then $[J]$ is the Padé approximant to $\langle h | f \rangle$ and $[\delta J]$ is the error of the Padé approximation. We define the differences

$$| \delta \psi_t \rangle = | \psi_t \rangle - | f \rangle \tag{6.8a}$$

and

$$\langle \delta \psi_t' | = \langle \psi_t' | - \langle f' |. \tag{6.8b}$$

From (6.3), (6.5), (6.6), and (6.8) we find, after cancellation, that

$$\delta[J] = - \langle \delta \psi_t' | 1 - \lambda A | \delta \psi_t \rangle. \tag{6.9}$$

This expression may be written symmetrically and be bounded by

$$|\delta[J]| \leq \frac{\| \langle \delta \psi_t' | (1 - \lambda A) \| \, \| (1 - \lambda A) | \delta \psi_t \rangle \|}{1 - |\lambda| \, \| A \|} \tag{6.10}$$

provided $|\lambda| < \| A \|^{-1}$.

We consider

$$(1 - \lambda A) | \delta \psi_t \rangle = (1 - \lambda A)(| \psi_t \rangle - | f \rangle)$$
$$= (1 - \lambda A) | \psi_t \rangle - | g \rangle \tag{6.11}$$

using (6.1) and (6.8a). From (5.24) and the preceding equations,

$$
|\psi_t\rangle = \frac{-1}{\det C}
\begin{vmatrix}
0 & c_0 & \cdots & c_{N-1} \\
|g\rangle & c_0 - \lambda c_1 & \cdots & c_{N-1} - \lambda c_N \\
A|g\rangle & c_1 - \lambda c_2 & \cdots & c_N - \lambda c_{N+1} \\
\vdots & \vdots & & \vdots \\
A^{N-1}|g\rangle & c_{N-1} - \lambda c_N & \cdots & c_{2N-2} - \lambda c_{2N-1}
\end{vmatrix},
$$

where

$$
\det C =
\begin{vmatrix}
c_0 - \lambda c_1 & \cdots & c_{N-1} - \lambda c_N \\
\vdots & & \vdots \\
c_{N-1} - \lambda c_N & \cdots & c_{2N-2} - \lambda c_{2N-1}
\end{vmatrix}
\tag{6.12}
$$

and $c_i = \langle h|A^i|g\rangle$ for $i = 0, 1, 2, 3, \ldots$. It follows from (6.1a) and (6.11) that

$$
(1 - \lambda A)|\psi_t\rangle - |g\rangle = (1 - \lambda A)|\delta\psi_t\rangle = \frac{-(-\lambda)^N}{\det C}|G\rangle
\tag{6.13}
$$

where

$$
|G\rangle =
\begin{vmatrix}
|g\rangle & c_0 & c_1 & \cdots & c_{N-1} \\
A|g\rangle & c_1 & c_2 & \cdots & c_N \\
\vdots & \vdots & \vdots & & \vdots \\
A^N|g\rangle & c_N & c_{N+1} & \cdots & c_{2N-1}
\end{vmatrix}.
\tag{6.14a}
$$

We define similarly

$$
\langle H| =
\begin{vmatrix}
\langle h| & c_0 & c_1 & \cdots & c_{N-1} \\
\langle h|A & c_1 & c_2 & \cdots & c_N \\
\vdots & \vdots & \vdots & & \vdots \\
\langle h|A^N & c_N & c_{N+1} & \cdots & c_{2N-1}
\end{vmatrix}.
\tag{6.14b}
$$

Noting that $\det C = Q^{[N-1/N]}(\lambda)$ in (6.12), it follows from (6.10), (6.12), (6.13), and (6.14) that

$$
|\delta[J]| \leq \frac{|\lambda|^{2N}\langle H|H\rangle^{1/2}\langle G|G\rangle^{1/2}}{(1 - \lambda\|A\|)Q^{[N-1/N]}(\lambda)^2}
\tag{6.15}
$$

provided that $|\lambda| < \|A\|^{-1}$. This formula is a bound on the error of the Padé approximants, which is calculated for

$$\delta[J] = \langle h | \chi_t \rangle - \langle h | f \rangle = \langle \chi_t' | g \rangle - \langle f' | g \rangle.$$

The construction of (6.15) does use additional information, such as values of $\langle g | (K^\dagger)^j K^k | g \rangle$, which are not used in the construction of the Padé approximants themselves. It is valid within the circle of convergence of the Neumann series of the integral equation. Most important, it is a formula comprising the accuracy-through-order concept, namely the factor $|\lambda|^{2N}$ in (6.15). It can be used to find bounds for Padé approximants to particular functions which may be represented by matrix elements of the solution of integral equations

$$\langle h | f \rangle = \sum_{i=0}^{\infty} c_i \lambda^i$$

within the circle of convergence. It is also easy to generalize it to refer to $[N+J-1/N-J]$ Padé approximants with $J = 0, 1, 2, \ldots, N$. Considerable development of this recent idea [Barnsley and Baker, 1976] is envisaged.

For further details of what may be achieved using methods similar to those of this section, we refer to Barnsley and Robinson [1974a–c] and Robinson and Barnsley [1979].

2.7 Single-Sign Potentials in Scattering Theory etc.

We consider in this section the problems of scattering from purely repulsive or purely attractive regular potentials in quantum mechanics. We can prove that the Padé approximants of the forward-scattering K-matrix and the partial-wave scattering amplitude converge. These results also hold for the Bethe–Salpeter equation in the one-particle-exchange approximation for equal-mass scattering. We start with potential theory and make the foregoing remarks more explicit.

The key requirement on the potential is that [Tani, 1966b; Masson, 1967a, b; Garibotti and Villani, 1969a, b]

$$V(r) = \lambda U(r) \qquad \text{with} \quad U(r) \geq 0 \quad \text{for all } r.$$

This enables $U^{1/2}(r)$ to be defined to be real and positive. In three-dimensional momentum space, it is represented by

$$U^{1/2}(p, q) = \int e^{-i(\mathbf{p}-\mathbf{q})\cdot\mathbf{r}} U^{1/2}(r) \, d\mathbf{r},$$

and in one dimension, in the lth partial wave,

$$U_l^{1/2}(p,q) = \int_0^\infty u_l(pr) U^{1/2}(r) u_l(qr) \, dq.$$

The K-matrix has been defined by (4.19):

$$\begin{aligned} K &= V + VPGK \\ &= V + KPGV \\ &= \lambda U + \lambda UPGK. \end{aligned} \tag{7.1}$$

This K-matrix has the Born–Neumann expansion

$$K = \lambda U + \lambda^2 UPGU + \lambda^3 UPGUPGU + \cdots. \tag{7.2}$$

(7.1) is a formal integral equation. It is easily made explicit in momentum space in one and three dimensions. The expansion (7.2) suggests that we focus attention on the symmetric kernel $U^{1/2}PGU^{1/2}$. This kernel is a real and symmetric $\mathfrak{L}^2$ kernel, and so Hilbert–Schmidt theory implies that

$$[U^{1/2}PGU^{1/2}] \doteq \sum_{i=0}^\infty \frac{\phi_i(\mathbf{p})\phi_i(\mathbf{q})}{\lambda_i}.$$

We have used the three-dimensional momentum-space interpretation; the λ_i are real, $\{\phi_i(\mathbf{p})\}$ are real orthogonal $\mathfrak{L}^2$ functions, and convergence is valid in the mean. We may now sum the series (7.2) after leaving the first two terms and obtain

$$K = \lambda U + \lambda^2 UPGU + \lambda^3 U^{1/2} \left[\sum_{i=0}^\infty \phi_i \frac{1}{\lambda_i(\lambda_i - \lambda)} \phi_i \right] U^{1/2}. \tag{7.3}$$

It follows that the residue of the ith pole of $K(p,p)/\lambda$ is

$$-\lambda_i \left[\int U^{1/2}(\mathbf{p},\mathbf{q})\phi_i(\mathbf{q}) \, d\mathbf{q} \right]^2,$$

which is real. Therefore $K(\mathbf{p},\mathbf{p})/\lambda$ has only poles with positive residues on the negative real axis and poles with negative residues on the positive real axis. An analogous result holds in one dimension for $K_l(p,p)/\lambda$. In this sense, we say that the K-matrix for forward scattering and the on-shell partial wave K-matrix are both Hamburger series, we deduce that the sequence of $[M-1/M]$ Padé approximants of both $K(p,p)/\lambda$ and $K_l(p,p)/\lambda$ converge. There is a slight ambiguity about how to define Padé approximants to (7.2), because the series may be interpreted as having a vanishing first term. We choose this interpretation, that the first term is zero, preserving the fundamental idea that the $[L/M]$ Padé approximant to (7.2) agrees with (7.2) to order λ^{L+M}. We recall from (4.23) that the on-shell partial-wave T-matrix is a homographic transform of the K-matrix,

$$T_l(k,k) = \frac{K_l(k,k)}{1 - ikK_l(k,k)} .$$

The usual accuracy-through-order proof shows that

$$[L/M]_{T_l(\lambda)} = [L/M]_{K_l(\lambda)} \{1 - ik[L/M]_{K_l(\lambda)}\}^{-1}$$

provided $L \leq M$. In particular, we note that Padé approximants of type $[M-1/M]$ for $K_l(\lambda)/\lambda$, suggested by the theory of Section 2.2 as the preferred choice, correspond to unitary $[M/M]$ approximants of the partial wave S-matrix. In conclusion, we have proved that the diagonal sequence of Padé approximants to the on-shell partial wave S-Matrix or T-matrix for single sign potential scattering converges and is unitary. It has also been proved from (7.3), [Garibotti and Villani, 1969b; Baker, 1975] that the numerators and denominators of the diagonal approximants of the S-matrix converge separately to Jost functions, as indicated by (4.52).

We obtain similar results for the Bethe–Salpeter equation by following a similar approach. We refer to the classic paper of Schwartz and Zemach [1966] for a fuller explanation of the scattering theory with the Bethe–Salpeter equation in the two-particle sector, using the Wick rotation, and we borrow their notation for the theory we need. The Wick rotation is a 90° rotation in the complex energy (p_0) plane which is valid for the Bethe–Salpeter equation below the three-particle production threshold and removes the two-particle propagator singularities from the equation. The wave function, which now involves a relative time coordinate, satisfies the integral equation

$$\psi(x) = e^{i\mathbf{k}\cdot\mathbf{r}} + \int d^4x' G(x,x')I(x')\psi(x'), \tag{7.4}$$

where $I(x')$ denotes the interaction, $G(x,x')$ is the equal-mass two-particle Green's function, and $x = (\mathbf{r}, t)$. The formulas are

$$G(x,x') = \int \frac{d^4p}{(2\pi)^4} \frac{e^{ip(x-x')}}{\left[\mathbf{p}^2 - (p_0+\omega)^2 + m^2 - i\varepsilon\right]\left[\mathbf{p}^2 - (p_0-\omega)^2 + m^2 - i\varepsilon\right]} \tag{7.5a}$$

$$= \frac{i}{16\pi\omega} \left\{ \frac{e^{ik|\mathbf{r}-\mathbf{r}'|}}{|\mathbf{r}-\mathbf{r}'|} - \left(\int_{-\infty}^{-\omega} + \int_{\omega}^{\infty} \right) \frac{d\beta}{\pi} e^{i\beta(t-t')} K_0(QR) \right\}, \tag{7.5b}$$

where

$$R=\sqrt{(r-r')^2-(t-t')^2}, \qquad Q=\sqrt{\beta^2-k^2},$$

$$\omega=\sqrt{k^2+m^2}, \quad \text{and} \quad |x|=\sqrt{r^2-t^2}.$$

The interaction is given by one-particle exchange:

$$I(x)=\frac{4\lambda\mu}{|x|}K_1(\mu|x|)\approx\frac{\lambda\sqrt{8\pi\mu}}{|x|^{\frac{3}{2}}}e^{-\mu|x|} \qquad \text{for} \quad |x|\to+\infty, \quad (7.6)$$

showing Yukawa exponential falloff for large $|x|$. A necessary constituent of the nonrelativistic and the relativistic analysis is a real Green's function, and we denote the standing-wave solutions associated with a real Green's function by a subscript r. The associated standing-wave function obeys

$$\psi_r(x)=e^{i\mathbf{k}\cdot\mathbf{r}}+\int d^4x'\,G_r(x,x')I(x')\psi_r(x'),$$

where

$$G_r(x,x')=\int\frac{d^4p}{(2\pi)^4}e^{ip(x-x')}$$

$$\times\left\{\left[\mathbf{p}^2-(p_0+\omega)^2+m^2-i\varepsilon\right]^{-1}\left[\mathbf{p}^2-(p_0-\omega)^2+m^2-i\varepsilon\right]^{-1}\right.$$

$$\left.+2\pi^2\delta^+\left[(p_0+\omega)^2-m^2\right]\delta^+\left[(p_0-\omega)^2-m^2\right]\right\} \qquad (7.7a)$$

$$=G(x,x')+\frac{i}{32\pi\omega|\mathbf{r}-\mathbf{r}'|}(e^{-ik|\mathbf{r}-\mathbf{r}'|}-e^{ik|\mathbf{r}-\mathbf{r}'|})$$

$$=\frac{i}{16\pi\omega}\left\{\frac{\cos k|\mathbf{r}-\mathbf{r}'|}{|\mathbf{r}-\mathbf{r}'|}-\left(\int_{-\infty}^{-\omega}+\int_{\omega}^{\infty}\right)\frac{d\beta}{\pi}e^{i\beta(t-t')}K_0(QR)\right\}.$$

$$(7.7b)$$

The asymptotic form, as $r\to\infty$ along the direction of $\mathbf{k}'$, of $\psi(x)$ follows from (7.4), (7.5b) and is

$$\psi(x)\sim e^{i\mathbf{k}\cdot\mathbf{r}}+\frac{ie^{ikr}}{16\pi\omega r}\int d^4x'\,e^{-i\mathbf{k}'\cdot x'}I(x')\psi(x').$$

This motivates the definition of the T-matrix, namely

$$T(\mathbf{k}',\mathbf{k})=\frac{i}{16\pi\omega}\int d^4x\,e^{-i\mathbf{k}'\cdot\mathbf{r}}I(x)\psi(x). \qquad (7.8a)$$

By analogy, Nuttall [1967] defines the R-matrix (the relativistic K-matrix or reaction matrix) as

$$R(\mathbf{k}',\mathbf{k})=\frac{i}{16\pi\omega}\int d^4x\, e^{-i\mathbf{k}'\cdot\mathbf{r}}I(x)\psi_r(x).\qquad(7.8b)$$

Equation (7.8a) gives the Born approximation to the T-matrix,

$$T^B(\mathbf{k}',\mathbf{k})=\frac{\pi\lambda/\omega}{(\mathbf{k}-\mathbf{k}')^2+\mu^2}.$$

We now make the Wick rotation, $t=\tau e^{-i\phi}$, in which ϕ increases from 0 to $\pi/2$ and τ is real. We define

$$\psi_r(\mathbf{r},t)=\phi_r(\mathbf{r},\tau)$$

and

$$I(\mathbf{r},t)=V(\mathbf{r},\tau).$$

It follows that $|x|=\sqrt{r^2+\tau^2}$ and $V(\mathbf{r},\tau)=4\lambda\mu K_1(\mu|x|)/|x|$. Thus we may define the positive quantity

$$U(\mathbf{r},\tau)=4\mu K_1(\mu|x|)|x|^{-1},$$

which has a well-defined square root. This overcomes the immediate problem of finding the analogue of the single-sign potential function. Next, we need a real integral equation with a symmetric kernel to determine $R(\mathbf{k},\mathbf{k}')$. This follows from (7.7b) using

$$H_r(x,x')=\frac{1}{16\pi\omega}\left\{\frac{\cos k|\mathbf{r}-\mathbf{r}'|}{|\mathbf{r}-\mathbf{r}'|}-\left(\int_{-\infty}^{-\omega}+\int_{\omega}^{\infty}\right)\frac{d\beta}{\pi}e^{\beta(\tau-\tau')}K_0(QR)\right\},$$

which is real. Hence we find

$$\phi_r=e^{i\mathbf{k}\cdot\mathbf{r}}+\lambda H_r U\phi_r,\qquad(7.9)$$

showing that ϕ_r obeys an integral equation with a real symmetric kernel. The inhomogeneous term of (7.9) is complex.

Equation (4.3) shows that the even-wave ($S, D, G,\dots$) components of $e^{i\mathbf{k}\cdot\mathbf{r}}$ are real, and that the odd-wave ($P, F, H,\dots$) components are pure imaginary. Equation (7.9) shows that ϕ_r shares this property. We express (7.8b) as

$$R(\mathbf{k}',\mathbf{k})=\frac{1}{16\pi\omega}\int d\mathbf{r}\,d\tau\, e^{-i\mathbf{k}'\cdot\mathbf{r}}V(\mathbf{r},\tau)\phi_r(\mathbf{r},\tau).\qquad(7.10)$$

The factor $e^{-i\mathbf{k}'\cdot\mathbf{r}}$ in (7.10) has an expansion of the form (4.3). We deduce that each partial-wave component of $R(\mathbf{k}',\mathbf{k})$ defined by (7.10) is real. Using a representation analogous to (7.3) and the discussion given previously, we see that the forward-scattering R-matrix, $R(\mathbf{k},\mathbf{k})/\lambda$, is a Hamburger series in the coupling strength. Similarly, we deduce that the on-shell partial-wave projections of $R(\mathbf{k}',\mathbf{k})/\lambda$ are Hamburger series in the coupling strength, with a convergent set of $[M-1/M]$ approximants. Further, the partial-wave T-matrix is given by the origin preserving homographic transform

$$T_l(k,k)=\frac{R_l(k,k)}{1-ikR_l(k,k)},$$

and so the diagonal sequence of Padé approximants to $T_l(k,k)$ also converges.

Partial-wave projections of (7.9) and (7.10) do not seem to have been used directly for solving the Bethe–Salpeter equation. These equations may be used with great success as ingredients for a Schwinger-type variational principle. For our purposes, they establish convergence of diagonal and subdiagonal sequences of Padé approximants to $T_l(k,k)$, and the momentum-space representation is used for calculations. This still leaves free the choices of whether to use a Wick rotation, subtraction methods, and the T or R matrix. The simplest conceptually is the unrotated, unsubtracted T-matrix equation, which explains the principles involved. Using (7.8a) and taking $k=(\mathbf{k},0)$, we define

$$T(k',k)=\frac{i}{16\pi\omega}\int d^4x\, e^{ik'x}I(x)\psi(x). \qquad (7.11)$$

Substituting (7.4) and (7.5a) into (7.11) gives

$$T(k',k)=V(k',k)$$

$$+\frac{16\pi\omega}{(2\pi)^4i}\int\frac{d^4q\,V(k',q)T(q,k)}{\left[\mathbf{q}^2-(q_0+\omega)^2+m^2-i\varepsilon\right]\left[\mathbf{q}^2-(q_0-\omega)^2+m^2-i\varepsilon\right]},$$

$$(7.12)$$

where

$$V(k',k)=\frac{i}{16\pi\omega}\int d^4x\, e^{i(k-k')x}I(x).$$

The fully off-shell extension is given by using k'' instead of the on-shell energy momentum k in (7.12), which then becomes

$$T(k';\omega;k'')=V(k',k'')$$

$$+\frac{\omega}{\pi^3i}\int\frac{d^4q\,V(k',q)T(q;\omega;k'')}{\left[\mathbf{q}^2-(q_0+\omega)^2+m^2-i\varepsilon\right]\left[\mathbf{q}^2-(q_0-\omega)^2+m^2-i\varepsilon\right]}.$$

$$(7.13)$$

The definitions

$$k'=(\mathbf{p}',\omega'), \quad p'=|\mathbf{p}'|, \quad k''=(\mathbf{p}'',\omega''), \quad p''=|\mathbf{p}''|$$

explain the partial-wave decomposition

$$T(k';\omega;k'')=\sum_{l=0}^{\infty}(2l+1)P_l(\cos\theta_{p',p''})t_l(\omega',p';\omega;\omega'',p''). \quad (7.14)$$

Following the method of (4.13), (4.14), and (4.15), we substitute (7.14) into (7.13), giving

$$t_l(\omega',p';\omega;\omega'',p'')=v_l(\omega',p';\omega'',p'')$$

$$+\frac{4\omega}{\pi^2 i}\int\frac{q^2\,dq\,dq_0\,v_l(\omega',p';q_0,q)t_l(q_0,q;\omega;\omega'',p'')}{\left[q^2-(q_0+\omega)^2+m^2-i\varepsilon\right]\left[q^2-(q_0-\omega)^2+m^2-i\varepsilon\right]}$$

$$(7.15)$$

This equation is the partial wave Bethe–Salpeter equation in momentum space, normalized so that the on-shell amplitude is

$$t_l\left(0,p;\sqrt{p^2+m^2};0,p\right)=\frac{e^{i\delta}\sin\delta}{k} \quad (7.16)$$

in the elastic region. The inhomogeneous term of (7.13) is given by

$$V(k',k)=\frac{\lambda}{\pi\omega}\frac{1}{(k'-k)^2+\mu^2-i\varepsilon},$$

and so the inhomogeneous term of (7.15) is given by

$$v_l(\omega',p';\omega'',p'')=\frac{\lambda}{\pi\omega}\frac{1}{2p'p''}Q_l\left(\frac{p'^2+p''^2+\mu^2-(\omega'-\omega'')^2-i\varepsilon}{2p'p''}\right).$$

$$(7.17)$$

Equations (7.15), (7.16), and (7.17) give the on-shell partial-wave t-matrix element, and we have already proved that the diagonal Padé sequence converges in the elastic region. We return to this in Section 2.8.

As a corollary to this section, concerned with exploitation of the kernel $U^{1/2}GU^{1/2}$, we include the results obtainable from the kernel $G^{1/2}UG^{1/2}$. No longer is U required to be single-signed, and the requirement falls on G.

For the Lippmann–Schwinger equation (7.19), we may take

$$G(q; k^2; q') = -\frac{2}{\pi} \frac{q^2}{k^2 - q^2 + i\varepsilon} \delta(q - q'),$$

which has a well-defined square root (as an operator)

$$G^{1/2}(q; k^2; q') = \sqrt{\frac{2}{\pi}} \frac{q}{(-k^2 + q^2)^{1/2}} \delta(q - q')$$

provided k^2 is negative.

For the Bethe–Salpeter equation, the Wick rotated kernel written in the form

$$H(x, x') = \frac{1}{16\pi^2\omega} \int_{-\omega}^{\omega} d\beta \, e^{\beta(\tau - \tau')} K_0(QR)$$

is real and positive provided $\omega < m$, which allows $H^{1/2}$ to be defined. We now write the T-matrix formally as $T = V + VGV + VG^{1/2}(G^{1/2}VG^{1/2} + (G^{1/2}VG^{1/2})^2 + \cdots)G^{1/2}V$, knowing that $G^{1/2}VG^{1/2}$ is a real symmetric kernel for $k^2 < 0$. Thus T/λ is a Hamburger series, and therefore the diagonal sequence of Padé approximants of T converges. The implications are that the (unphysical) forward-scattering T-matrix and partial-wave T-matrix have formal expansions, and the locations of poles of the Padé approximants of these are proved to determine the couplings for bound states of given energy of the system.

The rigorous results of this section can be extended to the case where the potential $V(r)$ does not have a single sign. The potential is decomposed as $V(r) = V_1(r) + V_2(r)$, where the whole interaction with $V_1(r)$ can be treated analytically (e.g. the square-well or Coulomb potential) and where $V_2(r)$ has a single sign. The eigenstates of the Hamiltonian including $V_1(r)$ [but not $V_2(r)$] are used as basis states, and then $V_2(r)$ is treated as a perturbation. Using this method, a distorted-wave formalism allows rigorous results about convergence of Padé approximants to the scattering amplitude to be established. We refer to Michalik [1970], Alder et al. [1973], Giraud et al. [1976], and Khalil [1977] for details.

So far in this section, we have considered cases in which the direct Padé methods for the Bethe–Salpeter and Lippmann–Schwinger equations are proved to converge. Numerical experience has shown that the direct Padé method is extremely successful when applied to the Bethe–Salpeter equation for nucleon–nucleon scattering, although no convergence proofs exist in this case. Plausible assumptions are made about the nature of the internucleon exchange forces, and the results of these calculations are most interesting as tests of the underlying theories [Gersten et al., 1976; Fleischer and Tjon, 1975].

2.8 Variational Padé Approximants

In Section 2.5, variational quantities such as $[E]$ in the Rayleigh–Ritz principle are minimized with respect to all possible variations of the trial functions. Indeed, we saw that the Padé method and the variational method give identical results using a particular basis for the trial functions. Naturally, this suggests that the Padé method may be improved by varying suitable parameters, such as the analogues of $\{\beta_{ij}\}$ described following (5.14) [Alabiso et al., 1970, 1971, 1972a].

In the bound-state and scattering problems of Sections 2.4 and 2.5, the full partial-wave Green's function $g(r, r', k^2)$ of (5.25) and its Padé approximant were expressed as functions of k^2, r, and r'. It is also convenient to use the momentum-space representation

$$\bar{g}(k', k'', k^2) = \int_0^\infty \int_0^\infty u_l(k'r)u_l(k''r)g(r, r', k^2)\,dr\,dr'. \qquad (8.1)$$

The point is that the T-matrix elements and their Padé approximants are expressed as functions of r, r' (or k', k'') as well as k^2. The poles of the exact Green's functions and the exact T-matrices occur at values of k which do not depend on r, r' (or k', k''), as follows from the theory of Section 2.2. Thus it is natural to evaluate the Padé approximants at values of r, r' (or k', k'') which are stationary points for $[I]$ or $[T]$ in (5.20) or (5.28). For example, the poles of the $[M-1/M]$ Padé approximant to $[T]$ depend on the off-shell momenta k' and k''. For scattering, the easiest method is to use on-shell values $k' = k'' = k$, but the best solution is to use values of k', k'' such that

$$\frac{\partial[T]}{\partial k'} = \frac{\partial[T]}{\partial k''} = 0. \qquad (8.2)$$

Figures 1 and 2 show that this leads to a marked improvement in the accuracy of low-order approximants. However, Alabiso, Butera, and Prosperi [1971] indicate that the proportional improvement is worse with higher-order approximants. Consequently, whenever calculational difficulties force one to use low-order Padé approximants, because only a few terms of the Maclaurin series are available, one would expect this variational device to be used whenever possible. There are, of course, ambiguities about what to do if the turning point is not unique, but it is usually possible to discriminate between a few discrete alternatives.

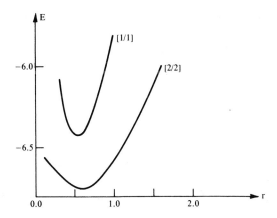

Figure 1. The energy of the first S-wave bound state, calculated from the [1/1] and [2/2] Padé approximants of the Green's function $g(r, r, k^2)$ in (5.25), against r, for the potential $V(r) = -29\exp(-r)$. [C. Alabiso et al., *Lett. Nuovo Cim.*, **3**, 831 (1970).]

For a further insight into the techniques likely to be successful in quantum field theories, we return to Equation (8.2). That use of the off-shell momentum as a variational parameter is an extraordinarily successful technique in scattering problems is an empirical fact. Nevertheless, the optimal value of this parameter is not given by the on-shell value of the momentum, even approximately. This deviation of empirical fact from physical intuition indicates a deficiency in the analysis, and it is the use of

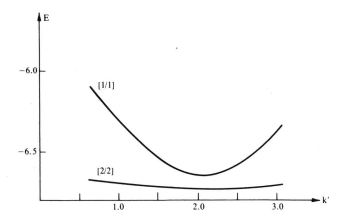

Figure 2. The energy of the first S-wave bound state, calculated from the [1/1] and [2/2] Padé approximants of the Green's function $\tilde{g}(k', k', k^2)$ in (7.1), against k', for the potential $V(r) = -29\exp(-r)$. [C. Alabiso et al., *Lett. Nuovo Cim.*, **3**, 831 (1970).]

matrix Padé approximants for the T-matrix that appears to offer the most natural solution to certain scattering problems.

The simplest matrix Padé method of calculating a scattering amplitude $\langle k|T|k \rangle$ consists of forming the power-series expansion of

$$\begin{pmatrix} \langle k|T|k \rangle & \langle k|T|k' \rangle \\ \langle k'|T|k \rangle & \langle k'|T|k' \rangle \end{pmatrix} = \sum_{i=1}^{\infty} \lambda^i t_i. \qquad (8.3)$$

Here each t_i is a 2×2 matrix; k is the on-shell momentum, and k' is the off-shell momentum, which may be used as the variational parameter. Equation (8.3) may be understood as an expression of four Born series of a particular partial-wave amplitude, expanded in the coupling strength λ. Other interpretations using a similar formalism are possible, such as that of a forward-scattering amplitude $\langle \mathbf{k}|T|\mathbf{k} \rangle$, etc. The formation of matrix Padé approximants of the series (8.3) enables the scattering amplitudes to be reconstructed; subsequently k' is varied until the reconstructed approximation of $\langle k|T|k \rangle$ is stationary.

We will show that this method unifies the techniques of variation of parameters of the trial wave function with the method of summation of the Born series, using the ideas of Section 2.5. Let $U(r)$ be the potential of unit strength, so that $V(r)=\lambda U(r)$ is the given potential. Our basic result is summarized by the result of Nuttall [1973]:

THEOREM 2.8.1. *Let* $g_1, g_2, \ldots, g_N$ *and* $h_1, h_2, \ldots, h_N$ *be scattering-state wave functions, denoted by vectors* $|\mathbf{g}\rangle$ *and* $\langle \mathbf{h}|$. *The matrix Padé method requires formation of an* $[M-1/M]$ *approximant of the matrix series*

$$T/\lambda = \sum_{i=0}^{\infty} \lambda^i \langle \mathbf{h}|(UG)^i U|\mathbf{g}\rangle .$$

If $1 \leqslant \alpha \leqslant N$, $1 \leqslant \beta \leqslant N$, *the value of* $T_{\alpha\beta} = \langle h_\alpha|T|g_\beta \rangle$ *derived by this method is the same as that derived from the Schwinger principle* (5.28) *for* $T_{\alpha\beta}$ *using trial functions which are linear combinations of*

$$g_1, g_2, \ldots, g_N, \; GUg_1, \ldots, GUg_N, \; (GU)^2 g_1, \ldots, (GU)^2 g_N, \ldots,$$
$$(GU)^{M-1} g_1, \ldots, (GU)^{M-1} g_N$$

and

$$h_1^*, h_2^*, \ldots, h_N^*, \; GUh_1^*, \ldots, GUh_N^*, \; (GU)^2 h_1^*, \ldots, (GU)^2 h_N^*, \ldots,$$
$$(GU)^{M-1} h_1^*, \ldots, (GU)^{M-1} h_N^*.$$

Proof. The method of proof is the same for arbitrary N as for $N=2$; for ease of presentation of the matrices we take $N=2$. Thus our trial wave functions are

$$|\psi_t\rangle = \sum_{i=0}^{M-1} d_i^{(1)}(GU)^i|g_1\rangle + \sum_{i=0}^{M-1} d_i^{(2)}(GU)^i|g_2\rangle$$

and

$$\langle\psi_t'| = \sum_{i=0}^{M-1} \tilde{d}_i^{(1)}\langle h_1|(UG)^i + \sum_{i=0}^{M-1} \tilde{d}_i^{(2)}\langle h_2|(UG)^i. \tag{8.4}$$

We consider the Schwinger bivariational functional

$$[T_{\alpha\beta}] = \langle\psi_t'|V|g_\beta\rangle + \langle h_\alpha|V|\psi_t\rangle - \langle\psi_t'|V-VGV|\psi_t\rangle, \tag{8.5}$$

where $\alpha,\beta = 1,2$. The functional is stationary when

$$\frac{\partial[T_{\alpha\beta}]}{\partial d_i^{(1)}} = \frac{\partial[T_{\alpha\beta}]}{\partial d_i^{(2)}} = \frac{\partial[T]_{\alpha\beta}}{\partial \tilde{d}_i^{(1)}} = \frac{\partial[T]_{\alpha\beta}}{\partial \tilde{d}_i^{(2)}} = 0$$

for $i=0,1,\ldots,M-1$. These lead to the equations

$$\langle h_\alpha|U(GU)^i|g_1\rangle - \langle\psi_t'|(U-\lambda UGU)(GU)^i|g_1\rangle = 0, \tag{8.6}$$

$$\langle h_\alpha|U(GU)^i|g_2\rangle - \langle\psi_t'|(U-\lambda UGU)(GU)^i|g_2\rangle = 0, \tag{8.7}$$

$$\langle h_1|(UG)^iU|g_\beta\rangle - \langle h_1|(UG)^i(U-\lambda UGU)|\psi_t\rangle = 0, \tag{8.8}$$

$$\langle h_2|(UG)^iU|g_\beta\rangle - \langle h_2|(UG)^i(U-\lambda UGU)|\psi_t\rangle = 0, \tag{8.9}$$

which hold for $i=0,1,\ldots,M-1$. Equations (8.6), (8.7) simplify to

$$T_{\alpha1}^{(i)} - \sum_{j=0}^{M-1} \tilde{d}_j^{(1)}\{T_{11}^{(i+j)} - \lambda T_{11}^{(i+j+1)}\}$$

$$- \sum_{j=0}^{M-1} \tilde{d}_j^{(2)}\{T_{21}^{(i+j)} - \lambda T_{21}^{(i+j+1)}\} = 0,$$

$$T_{\alpha2}^{(i)} - \sum_{j=0}^{M-1} \tilde{d}_j^{(1)}\{T_{12}^{(i+j)} - \lambda T_{12}^{(i+j+1)}\}$$

$$- \sum_{j=0}^{M-1} \tilde{d}_j^{(2)}\{T_{22}^{(i+j)} - \lambda T_{22}^{(i+j+1)}\} = 0,$$

which are $2M$ equations for $\tilde{d}_j^{(1)}, \tilde{d}_j^{(2)}, j=0,1,\ldots, M-1$. They can be written conveniently in block matrix form as

$$
\begin{pmatrix} T_{11}-\lambda T_{11}^{(+)} & T_{21}-\lambda T_{21}^{(+)} \\ T_{12}-\lambda T_{12}^{(+)} & T_{22}-\lambda T_{22}^{(+)} \end{pmatrix} \begin{pmatrix} \tilde{\mathbf{d}}^{(1)} \\ \tilde{\mathbf{d}}^{(2)} \end{pmatrix} = \begin{pmatrix} T_{\alpha 1} \\ T_{\alpha 2} \end{pmatrix},
$$

where the elements of the $M\times M$ block matrices are

$$
\left(T_{\alpha\beta}\right)_{ij}=T_{\alpha\beta}^{(i+j)} \text{ and } \left(T_{\alpha\beta}^{(+)}\right)_{ij}=T_{\alpha\beta}^{(i+j+1)}, \qquad i,j=0,\ldots, M-1.
$$

Hence

$$
(\tilde{\mathbf{d}}^{(1)}\ \tilde{\mathbf{d}}^{(2)})=\begin{pmatrix} T_{\alpha 1} & T_{\alpha 2} \end{pmatrix} \begin{pmatrix} T_{11}-\lambda T_{11}^{(+)} & T_{12}-\lambda T_{12}^{(+)} \\ T_{21}-\lambda T_{21}^{(+)} & T_{22}-\lambda T_{22}^{(+)} \end{pmatrix}^{-1}. \tag{8.10}
$$

Similarly we may derive from (8.8), (8.9) that

$$
\begin{pmatrix} \mathbf{d}^{(1)} \\ \mathbf{d}^{(2)} \end{pmatrix} = \begin{pmatrix} T_{11}-\lambda T_{11}^{(+)} & T_{12}-\lambda T_{12}^{(+)} \\ T_{21}-\lambda T_{21}^{(+)} & T_{22}-\lambda T_{22}^{(+)} \end{pmatrix}^{-1} \begin{pmatrix} T_{1\beta} \\ T_{2\beta} \end{pmatrix}. \tag{8.11}
$$

Substituting (8.10), (8.11) in (8.4), (8.5), the stationary value of $[T_{\alpha\beta}]$ is given by

$$
[T_{\alpha\beta}]_{st}=\begin{pmatrix} T_{\alpha 1} & T_{\alpha 2} \end{pmatrix} \begin{pmatrix} T_{11}-\lambda T_{11}^{(+)} & T_{12}-\lambda T_{12}^{(+)} \\ T_{21}-\lambda T_{21}^{(+)} & T_{22}-\lambda T_{22}^{(+)} \end{pmatrix}^{-1} \begin{pmatrix} T_{1\beta} \\ T_{2\beta} \end{pmatrix}. \tag{8.12}
$$

This is the Nuttall compact form of the 2×2 matrix Padé approximant. The $N\times N$ generalized form of this formula is entirely obvious, and the result for general N follows by this proof.

Nuttall's theorem establishes the N-dimensional $[M-1/M]$ matrix Padé method as a variational method in a $2MN$-dimensional space, thereby accounting for the remarkable accuracy of the method.

It is interesting to note that the $[0/1]$ two-dimensional variational matrix Padé method is exact for scattering from a square-well potential [Fratamico et al., 1976; Graves-Morris, 1978b]:

$$
g_1=h_1^* \propto j_l(kr)
$$

for the on-shell wave function [see (4.31a)], and

$$
g_2=h_2^* \propto j_l(qr)
$$

for the off-shell wave function, depending on the variational parameter q. Let $V(r) = \lambda\theta(b-r)$ be a square-well potential of strength λ and range b. With units in which $\hbar = 2m = 1$, the momentum q inside the well is given by

$$q^2 = k^2 - \lambda. \tag{8.13}$$

With this value of q, $g_2(q)$ is the exact wave function within the well. However, (8.5) is independent of the wave function outside the well, and consequently g_2 is effectively the exact wave function if q is given by (8.13). The Schwinger principle asserts that $[T_{\alpha\beta}]$ is the exact scattering amplitude and that $|\psi_t\rangle$ is the exact wave function if the exact wave function is a linear combination of the basis wave functions. Therefore $[T_{\alpha\beta}]$ has a turning point and takes the exact value of the scattering amplitude when q is given by (8.13). Note that (8.13) does not necessarily give the only turning point of $[T_{\alpha\beta}]$, and that the order of Padé approximation is immaterial: the lowest-order $[0/1]$ approximant suffices for an exact result.

The moral of this analysis is that a variational 2×2 matrix Padé method is probably as efficient a method as any for scattering from a single Yukawa-like potential; what is best for the realistic, sign changing nuclear potentials is not yet known.

As a prelude to the use of variational Padé approximants in quantum field theory, the Bethe–Salpeter equation was tackled [Alabiso et al., 1972b]. Equation (7.15) shows that ω', p', ω'', and p'' are the four available variational parameters. This cornucopia is grasped and controlled by the empirical observation that $\omega' = \omega'' = 0$ is usually the optimal choice. In short, on-shell energy variables should be used. The authors chose to work with a single symmetric variational parameter by taking $p' = p''$ in the Wick rotated, once subtracted Bethe–Salpeter R-matrix equation. Their results confirm the view that for good accuracy in low-order Padé approximation, a variational technique is essential. We show in Figure 3 their results for the first S-wave bound state, at $s = 1$, corresponding to $\omega = -\frac{1}{2}$ with $m = \mu = 1$ [see (7.5) and (7.6)]. The figure shows the values of λ needed for this bound state versus the variational momentum p'. There is a clear turning point near $p' = 1.25$, and the importance of using the variational principle to select this value is very evident.

In practice, variational methods have been very successful in conjunction with the solution of the Bethe–Salpeter equation for nucleon–nucleon scattering. A variety of tricks have been used to reduce the problems to manageable proportions, and we refer to Fleischer et al. [1973], Fleischer and Tjon [1980], and Bessis and Turchetti [1977] for details.

For further details on matrix Padé approximants and variational techniques, we refer to Turchetti [1976], Benofy and Gammel [1977], Benofy et al. [1976], Bessis et al. [1977], Mery [1977], and Pindor [1979b].

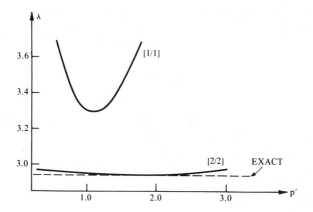

Figure 3. The coupling constant λ required to give a bound state at $s=1$, against the off-shell momentum p' for the [1/1] and [2/2] Padé approximants to the S-wave Bethe–Salpeter R-matrix.

2.9 Singular Potentials

A remarkable advantage of Padé approximants is their ability to treat the problems of singular potentials. A potential such as $V(r)=\lambda r^{-4}$ is so strong at short range that any negative value of λ causes bound states of infinite binding energy. This may be deduced from the Rayleigh–Ritz principle (5.2). Normally, only positive values of λ are considered to be physical. From a mathematical point of view, a Frobenius expansion of the radial Schrödinger equation

$$\psi_l(r)=r^\alpha\left(a_0+a_1r+a_2r^2+\cdots\right)$$

leads to an indicial equation for α which cannot be satisfied. Potentials more divergent at the origin than $V(r)=\lambda r^{-2}$ are called singular potentials. Schrödinger's equation still has scattering solutions corresponding to scattering from a repulsive singular potential, and it turns out that $\psi_l(r)$ has an essential singularity at $r=0$ in this case. The review of Frank et al. [1971] contains an exhaustive account of the best-known examples of singular potentials and their properties. The Padé method consists of regularizing the potentials with a cutoff [Garibotti et al., 1970]. The general method is quite clear from a specific example, and so we consider the potential

$$V(r)=\lambda r^{-4} \tag{9.1}$$

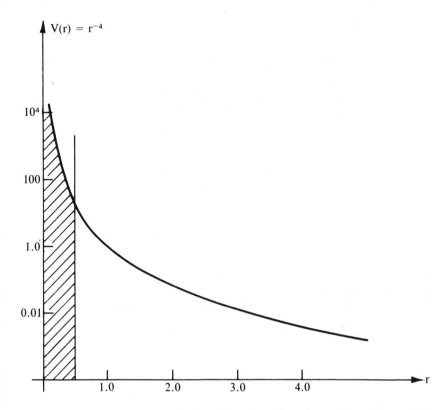

Figure 1. The potential $V(r)=r^{-4}$ plotted *logarithmically* against r. A cutoff at $r=0.5$ is shown.

and define a cut-off potential

$$V_\varepsilon(r)=\lambda r^{-4}\theta(r-\varepsilon) \qquad (9.2a)$$

$$=\lambda r^{-4}, \qquad r>\varepsilon$$
$$=0, \qquad r\leqslant\varepsilon \qquad (9.2b)$$

We are going to use the Jost method, and so the equation for the Jost functions, (4.51),

$$f(k,r)=e^{ikr}-\frac{1}{k}\int_r^\infty \sin k(r-r')V_\varepsilon(r')f(k,r')\,dr', \qquad (9.3)$$

is a Volterra integral equation with an $\mathcal{L}^2$ kernel. If the potential were not cut off, the kernel would not be $\mathcal{L}^2$. We will use an approximation valid for small k. It follows that

$$f(0,r)=1+\int_r^\infty (r'-r)V_\varepsilon(r')f(0,r')\,dr', \qquad (9.4a)$$

$$\frac{\partial f}{\partial k}(0,r)=ir+\int_r^\infty (r'-r)V_\varepsilon(r')\frac{\partial f}{\partial k}(0,r)\,dr'. \qquad (9.4b)$$

Define

$$f_A(k,r)=f(0,r)+k\frac{\partial f}{\partial k}(0,r) \qquad (9.5)$$

Then

$$f_A(k,r)=1+ikr+\int_r^\infty (r'-r)V_\varepsilon(r')f_A(k,r')\,dr'.$$

We solve this equation iteratively for $r>\varepsilon$. (Of course, this calculation is easy for our particular example, but even in more complicated situations it is quite straightforward to provide a numerical iterative solution.) We find

$$f_A(k,r)=1+ikr+\lambda\left(\frac{r^{-2}}{3!}+\frac{r^{-1}}{2!}ik\right)+\lambda^2\left(\frac{r^{-4}}{5!}+\frac{r^{-3}}{4!}ik\right)+\cdots$$

$$=\frac{r}{\sqrt{\lambda}}\sinh\frac{\sqrt{\lambda}}{r}+ikr\cosh\frac{\sqrt{\lambda}}{r} \qquad \text{for} \quad r>\varepsilon.$$

We need to take $r=0$ for the Jost function; the easiest method of doing this is to use (9.4) to provide a formula for $f_A(k,r)$ valid at $r=0$:

$$f_A(k,0)=1+\int_\varepsilon^\infty r^{-2}\left(\sqrt{\lambda}\sinh\frac{\sqrt{\lambda}}{r}+ik\lambda\cosh\frac{\sqrt{\lambda}}{r}\right)dr$$

$$=\cosh\frac{\sqrt{\lambda}}{\varepsilon}+ik\sqrt{\lambda}\sinh\frac{\sqrt{\lambda}}{\varepsilon}.$$

Hence

$$f_A(-k,0)=\cosh\frac{\sqrt{\lambda}}{\varepsilon}-ik\sqrt{\lambda}\sinh\frac{\sqrt{\lambda}}{\varepsilon},$$

and from (4.52),

$$\exp(2i\delta)=\frac{1-ik\sqrt{\lambda}\tanh(\sqrt{\lambda}/\varepsilon)}{1+ik\sqrt{\lambda}\tanh(\sqrt{\lambda}/\varepsilon)}+O(k^2).$$

Our convention is that $\lambda>0$ gives a positive, repulsive potential and normally a negative phase shift. But, for notational convenience in this section only, we use the nuclear-physics convention that the scattering length a is positive in this situation. Therefore the scattering length, which depends on the coupling strength λ and the cutoff ε, is given by

$$a(\varepsilon, \lambda) = \sqrt{\lambda} \tanh \frac{\sqrt{\lambda}}{\varepsilon}. \tag{9.6}$$

At this point we must assume that the scattering length $a(\varepsilon, \lambda)$ caused by the potential $V(r)$ is the limit as $\varepsilon \to 0$ of the cut-off potentials $V_\varepsilon(r)$. Then we deduce that $a(\lambda) = \lambda^{1/2}$. Let us note the following important features of the cutoff solution, which all follow (9.6):

(i) $\dfrac{a(\varepsilon, \lambda)}{\lambda} = \dfrac{1}{\varepsilon} - \dfrac{1}{3}\dfrac{\lambda}{\varepsilon^3} + \dfrac{2}{15}\dfrac{\lambda^2}{\varepsilon^5} - \cdots = \dfrac{1}{\sqrt{\lambda}} \tanh \dfrac{\sqrt{\lambda}}{\varepsilon}$. This is a Stieltjes series which is regular at $\lambda = 0$, and has poles at $\lambda = -(\varepsilon n \pi / 2)^2$ for n odd and integral.

(ii) $\displaystyle\lim_{\varepsilon \downarrow 0} \left(\frac{\partial}{\partial \varepsilon}\right)^n a(\varepsilon, \lambda) = 0$ for all $n > 0$.

(iii) $a(\varepsilon, \lambda) < a(\lambda)$ for $\varepsilon > 0$.

(iv) $[N-1/N]_{\lambda^{-1}a(\varepsilon, \lambda)} = \dfrac{1}{\varepsilon} \dfrac{p_{N-1}(\lambda/\varepsilon^2)}{q_N(\lambda/\varepsilon^2)}$, where p_{N-1} and q_N are polynomials of degrees $N-1$ as N respectively.

Let us define

$$a_N(\varepsilon) = [N/N]_{a(\varepsilon, \lambda)} = \lambda[N-1/N]_{\lambda^{-1}a(\varepsilon, \lambda)}.$$

Then we find the surprising result that when the cutoff is zero or infinity,

$$a_N(0) = a_N(\infty) = 0.$$

These properties are shown in Figure 2, which is a plot of $a_N(\varepsilon, \lambda)/\lambda^{1/2}$ against $\varepsilon/\lambda^{1/2}$. The value of the scattering length with no cutoff, $\lambda^{1/2}$, is shown as a straight line with ordinate one, corresponding to the limiting values of $a_N(\varepsilon, \lambda)/\lambda^{1/2}$.

We are now able to understand what happens in the general case of an arbitrary singular repulsive potential at zero energy. The arguments of Section 2.7 show that $\lambda^{-1}a(\varepsilon, \lambda)$ is a Stieltjes series. Using either Jost solutions or the Lippmann–Schwinger equation, we can construct $a(\varepsilon, \lambda)$ as a power series in λ. We can then form Padé approximants to the series, and use the theorems of Part I, Chapter 5 to prove that

$$[N/N]_{\lambda^{-1}a(\varepsilon, \lambda)} \geqslant \frac{a(\varepsilon, \lambda)}{\lambda} \geqslant [N-1/N]_{\lambda^{-1}a(\varepsilon, \lambda)}.$$

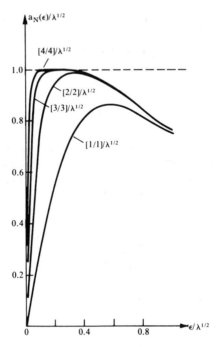

Figure 2. $[N/N]$ Padé approximants to the scattering length for the regularized potential $V(r)=\lambda r^{-4}\theta(r-\varepsilon)$. From P. R. Graves-Morris (1973).

Hence

$$a_N(\varepsilon,\lambda)\equiv[N/N]_{a(\varepsilon,\lambda)}\leqslant a(\varepsilon,\lambda)$$

and

$$\lim_{N\to\infty} a_N(\varepsilon,\lambda)=a(\varepsilon,\lambda).$$

The whole purpose of our work is to find

$$a(\lambda)=\lim_{\varepsilon\to 0} a(\varepsilon,\lambda)=\lim_{\varepsilon\to 0}\lim_{N\to\infty} a_N(\varepsilon,\lambda). \tag{9.7}$$

The example $V(r)=\lambda r^{-4}$ shows that the limits cannot be interchanged, because $a_N(0,\lambda)=0$. In fact, $a_N(0,\lambda)=0$ is a general result for all singular potentials, and the combined result

$$a_N(0,\lambda)=a_N(\infty,\lambda)=0 \tag{9.8}$$

follows by power counting. To understand the limits of (9.7) we exhibit the

Table 1. $[N/N]$ Approximants of Scattering Lengths of Cut-Off Potentials

$a_1(\varepsilon_1)$	$a_1(\varepsilon_2)$	$a_1(\varepsilon_3)$	$\cdots$	$\rightarrow$	0
$a_2(\varepsilon_1)$	$a_2(\varepsilon_2)$	$a_2(\varepsilon_3)$	$\cdots$	$\rightarrow$	0
$a_3(\varepsilon_1)$	$a_2(\varepsilon_2)$	$a_3(\varepsilon_3)$	$\cdots$	$\rightarrow$	0
$\vdots$	$\vdots$	$\vdots$			
$\downarrow$	$\downarrow$	$\downarrow$			
$a(\varepsilon_1)$	$a(\varepsilon_2)$	$a(\varepsilon_3)$	$\cdots$	$\rightarrow$	$a(0)$

entries in Table 1. We suppress the variable λ for conciseness. Table 1 shows $a_N(\varepsilon_i)$, the $[N/N]$ Padé approximant to the scattering length for a singular potential cut off at $r=\varepsilon_i$. Each column has a limit $a(\varepsilon_i)$ as $N\rightarrow\infty$, and each row has the limit zero as $\varepsilon_i\rightarrow 0$.

The mathematics indicates that we should extrapolate each column to its limit $a(\varepsilon_i)$ and deduce the scattering length $a(0)$ as $\varepsilon_i\rightarrow 0$. Even if this is feasible, it is an inefficient procedure, and there is a much better method.

Because all the entries in Table 1 are positive, (9.8) shows that the maximum entry of each row occurs at a finite, non-zero value of ε. Thus $a_N(\varepsilon,\lambda)$ has a maximum value at $\varepsilon=\bar{\varepsilon}(N)$, and if there are several maxima, $\bar{\varepsilon}(N)$ is the largest of these. Then we find that

$$a(0)= \lim_{N\rightarrow\infty} a_N(\bar{\varepsilon}(N)), \qquad (9.9)$$

which is a stable procedure.

To prove this correct, take $\eta>0$, and we prove that, for all $N>N_0(\eta)$,

$$|a_N(\bar{\varepsilon}(N))-a(0)|<\eta.$$

The proof begins by noting the existence of a cutoff ε, depending on η, for which, by our continuity hypothesis,

$$|a(\varepsilon)-a(0)|<\eta/2$$

and then noting the order $N_0(\eta)$ for which

$$|a_N(\varepsilon)-a(\varepsilon)|<\eta/2$$

for all $N>N_0(\eta)$. Since we chose $\bar{\varepsilon}(N)$ so that

$$a_N(\bar{\varepsilon}(N))\geq a_N(\varepsilon_i)$$

for all ε_i in row N of the table, it follows that

$$|a_N(\bar{\varepsilon}(N)) - a(0)| \leqslant |a_N(\varepsilon) - a(0)|$$
$$\leqslant |a_N(\varepsilon) - a(\varepsilon)| + |a(\varepsilon) - a(0)|$$
$$< \eta$$

provided $N > N_0(\eta)$. Hence (9.9) is proved.

This argument formally proves the convergence of the method, as summarized by (9.9). The fact that we use the maximum entry in each row shows that the process is efficient. Figure 2 demonstrates the efficiency of the method for our particular example $V(r) = \lambda r^{-4}$.

The possibilities of application of this technique in divergent theories would seem to be immense. Some applications are briefly reviewed by Graves-Morris [1973], and a full discussion of convergence is given by Bessis et al. [1974].

CHAPTER 3

Connection with Numerical Analysis

3.1 Gaussian Quadrature

There is an intimate formal connection between Gaussian quadrature and Padé approximation, which we will describe in this section. The basic quadrature problem is usually posed as the problem of finding $I\{g\}$, a formula for the numerical integration of an arbitrary function $g(x)$ over a Stieltjes measure $d\phi(u)$ on the interval (a, b). For Gaussian quadrature, $I(g)$ is a weighted mean given by

$$I\{g\} = \sum_{i=1}^{m} g(u_i)w_i = \int_a^b g(u)\,d\phi(u) + e_m\{g\}. \tag{1.1}$$

This is an m-point formula, using points u_i and weights w_i, $i = 1, 2, \ldots, m$; $e_m\{g\}$ is the quadrature error, which should be small. To define the points and weights, which are $2m$ parameters, the formula (1.1) is required to integrate the $2m$ monomials $u^0, u^1, u^2, \ldots, u^{2m-1}$ exactly, which means that

$$\int_a^b u^j\,d\phi(u) = \sum_{i=1}^{m} (u_i)^j w_i, \qquad j = 0, 1, 2, \ldots, 2m-1. \tag{1.2}$$

With the points and weights determined by (1.2), we see that the linear form of (1.1) integrates any polynomial of maximum degree $2m-1$ exactly. In particular, it integrates the best weighted polynomial approximation of maximum order $2m-1$ to $g(u)$ exactly, and so it is in this sense that $e_m\{g\}$ is small.

ENCYCLOPEDIA OF MATHEMATICS and Its Applications, Gian-Carlo Rota (ed.). Vol. 14: George A. Baker, Jr., and Peter R. Graves-Morris, Padé Approximants, Part II: Extensions and Applications   ISBN 0-201-13513-2

Our starting point is the observation that

$$f(z)=\int_a^b \frac{d\phi(u)}{1+zu}=\sum_{i=1}^m \frac{w_i}{1+zu_i}+O(z^{2m}) \tag{1.3}$$

represents the formation of the $[m-1/m]$ Padé approximant to $f(z)$. By the definition (I.5.6.1), $f(z)$ is a Hamburger function and so (Part I, Section 5.5) $u_i \in [a,b]$, $w_i>0$, $i=1,2,\ldots,m$, and of course the Padé approximant exists and is nondegenerate. Equating coefficients of $(-z)^j$ in (1.3) yields (1.2) and provides a complete justification of (1.2). Thus we see that the poles of the $[m-1/m]$ Padé approximants of $f(z)$ are the quadrature points u_i and the residues (w_i/u_i) determine the weights w_i.

The quadrature error is given implicitly by the characteristic function for the quadrature formula defined [Takehasi and Mori, 1971] by

$$E_m(z)=f(z)-[m-1/m]$$

$$=\int_a^b \frac{d\phi(u)}{1+zu}-\frac{P^{[m-1/m]}(z)}{Q^{[m-1/m]}(z)}. \tag{1.4}$$

Following the methods of Part I, Sections 5.3–5.6 we introduce the variable $w=-z^{-1}$. Let $\pi_0(w)=1$ and then (I.5.3.20) defines polynomials $\pi_m(w)$ by

$$\pi_m(w)=w^m Q^{[m-1/m]}(-1/w), \qquad m=1,2,\ldots .$$

Recall (I.5.3.18), which shows that $\pi_m(w)$ are orthogonal polynomials over the measure $d\phi(u)$ on $[a,b]$ if $f(z)$ is a Stieltjes function and (see Part I, page 210) that the result generalizes to Hamburger functions also. From (1.4), we find that

$$E_m(z)=\frac{1}{Q^{[m-1/m]}(z)}\left\{\int_a^b \frac{w^{-m}\pi_m(w)\,d\phi(u)}{1+zu}-P^{[m-1/m]}(z)\right\}$$

$$=\frac{1}{Q^{[m-1/m]}(z)}\left\{\int_a^b w^{-m+1}\frac{\pi_m(w)-\pi_m(u)}{w-u}d\phi(u)\right.$$

$$\left.+\int_a^b \frac{w^{-m+1}\pi_m(u)}{w-u}d\phi(u)-P^{[m-1/m]}(z)\right\}.$$

By using the accuracy-through-order condition for $E_m(z)$, we deduce that

$$E_m(z)=\frac{(-z)^m}{Q^{[m-1/m]}(z)}\int_a^b \frac{\pi_m(u)}{1+uz}d\phi(u). \tag{1.5}$$

By writing $(1+uz)^{-1}=1-uz+u^2z^2+\cdots+(-)^m(uz)^m/(1+uz)$, and using orthogonality, (1.5) leads to

$$E_m(z)=\frac{z^{2m}}{Q^{[m-1/m]}(z)}\int_a^b\frac{u^m\pi_m(u)}{1+uz}\,d\phi(u),\qquad(1.6)$$

which shows explicitly that $E_m(z)=O(z^{2m})$. Further analysis of (1.6) gives

$$E_m(z)=\left\{\frac{z^m}{Q^{[m-1/m]}(z)}\right\}^2\int_a^b\frac{w^{-m}\pi_m(w)\pi_m(u)}{1+uz}u^m\,d\phi(u)$$

$$=\left\{\frac{z^m}{Q^{[m-1/m]}(z)}\right\}^2\int_a^b\frac{w^{1-m}\pi_m(w)\pi_m(u)}{w-u}u^m\,d\phi(u).\qquad(1.7)$$

Noting that

$$\frac{(u/w)^m\pi_m(w)-\pi_m(u)}{w-u}$$

is a polynomial in u of degree $m-1$, we find that (1.7) becomes

$$E_m(z)=\left\{\frac{z^m}{Q^{[m-1/m]}(z)}\right\}^2\int_a^b\frac{\{\pi_m(u)\}^2}{1+uz}\,d\phi(u).\qquad(1.8)$$

Equations (1.5)–(1.8) give different representations of the characteristic function. The coefficient of z^k in $E_m(z)$ determines the quadrature error with a Gaussian quadrature formula. In particular, for the least nonzero coefficient,

$$e_m(u^{2m})=\frac{1}{[C(m-1/m)]^2}\int_a^b\{\pi_m(u)\}^2\,d\phi(u).\qquad(1.9)$$

Notice that $\pi_m(u)/C(m-1/m)$ is the mth orthogonal polynomial with leading coefficient unity, as expected for the optimal quadrature formula. For further details, we refer to Allen et al. [1974], Karlsson and von Sydow [1976], and Brezinski [1977].

Next, we turn our attention to the matter of extrapolating quadrature estimates to give accurate answers for numerical integration. Various estimates are made for the value of a definite integral using $m_1, m_2, m_3,\ldots$ point quadrature rules, and these results are extrapolated to infinite precision. Let us assume that the integral

$$I\{f\}\equiv\int_0^1 f(x)\,dx\qquad(1.10)$$

exists. Estimates of its value are made using the m-fold midpoint trapezoidal rule,

$$Q^m\{f\} \equiv \frac{1}{m} \sum_{j=1}^{m} f\left(\frac{2j-1}{2m}\right), \qquad (1.11)$$

in which the m points are equally spaced with spacing $h = 1/m$. The assumed existence of $I\{f\}$ implies the existence of a function $\phi(h)$ which is continuous on $0 \leqslant h \leqslant 1$ for which

$$Q^{(m)}\{f\} = \phi(1/m), \qquad m = 1,2,3,\dots, \qquad (1.12)$$

and

$$I\{f\} = \phi(0).$$

For example, three estimates of the value of (1.10) may be made using

$$m_1, \qquad m_2 = 2m_1, \quad \text{and} \quad m_3 = 2m_2 \qquad (1.13)$$

quadrature points. We set

$$\phi(1/m_i) = Q^{(m_i)}\{f\}, \qquad i = 1,2,3. \qquad (1.14)$$

The fundamental problem is that of extrapolating these three values to $h = 0$ to obtain a better estimate of $I\{f\}$. The extrapolation may be performed, for example, using a quadratic polynomial in h, by using a [1/1]-type rational fraction in h, or by using Aitken's Δ^2 method. We emphasize that one may choose m_1, m_2, and m_3 differently from (1.13) and that there is no reason in general to restrict oneself to precisely three quadrature estimates. To answer the fundamental question of which type of extrapolation is likely to be most effective, we need estimates of the error. The error is given by the Euler–Maclaurin sum formula [Lyness and Ninham, 1967],

$$Q^{(m)}\{f\} - I\{f\} = \sum_{n=1}^{\infty} (-1)^n \frac{\zeta(2n)\left[2 - \left(\frac{1}{4}\right)^{n-1}\right]}{(2\pi m)^{2n}} \left[f^{(2n-1)}(1) - f^{(2n-1)}(0)\right], \qquad (1.15)$$

which is valid whenever $f \in C^\infty[0,1]$ and the right-hand side is convergent. The factor $\zeta(2n)$ occurring on the right-hand side is the Riemann zeta function [Abramowitz and Stegun, 1964, Chapter 23], and $\zeta(2n) \to 1$ as $n \to \infty$. If the derivatives $f^{(2n-1)}$ grow faster than $(2\pi m)^{2n}$, then divergence of the right-hand side of (1.15) is expected. This divergence occurs, for example, if f is the logarithmic function. The closed Euler–Maclaurin sum rule, equivalent to the open rule (1.15), can be used to derive the asymptotic expansion of the gamma function (see Part I, Section 5.5). The interrelations of these results are extensively discussed by Hardy [1956, Chapter 13] in this context. In such cases, (1.15) is to be understood as an asymptotic equality, which is to say that if the sum on the right-hand side is truncated at a finite number of terms, then (1.15) is valid as a finite power series in m^{-1} in the limit $m \to \infty$. This procedure may be very effective in practice, as may be seen by taking just two terms of (1.15),

$$Q^{(m)}\{f\} - I\{f\} = -\frac{1}{24m^2}\left[f^{(1)}(1) - f^{(1)}(0)\right]$$

$$+ \frac{7}{7560m^4}\left[f^{(3)}(1) - f^{(3)}(0)\right] + \cdots.$$

We refer to Hardy [1956] and Ninham and Lyness [1969] for further details of these asymptotic series. We consider a specific example.

$$f(x) = x^{-3/4}(1-x)^{-1/2}h(x) \tag{1.16}$$

in which $h(x) \in C^{\infty}[0,1]$. For this case, one finds [Lyness and Ninham, 1967] that

$$Q^{(m)}\{f\} - I\{f\} = \frac{a_0}{m^{1/4}} + \frac{a_1}{m^{5/4}} + \frac{a_2}{m^{9/4}} + O(m^{-13/4})$$

$$+ \frac{b_0}{m^{1/2}} + \frac{b_1}{m^{3/2}} + \frac{b_2}{m^{5/2}} + O(m^{-7/2}). \tag{1.17}$$

The implication of (1.17) is that $\phi'(h) = \infty$ at $h = 0$, even though $\lim_{h \to 0} \phi(h)$ is well defined. This circumstance strongly suggests that some form of rational interpolation is likely to be more effective than linear interpolatory methods in this case, in which (1.17) is effectively a rational function of $m^{-1/4}$. It is a curious serendipity that Padé and rational interpolation methods seem to be reliable and efficient whenever linear methods of quadrature extrapolation converge unacceptably slowly. We refer to Joyce [1971] for a comprehensive review of both linear and nonlinear methods of acceleration of convergence of quadrature estimates. For further details of the Padé methods, we refer to Chisholm et al. [1972], Genz [1972, 1973, 1974] and Werner and Wuytack [1978].

Exercise 1. Find the form of (1.15) applicable to quadrature on an interval $[a, b]$.

Exercise 2. What changes in the analysis of (1.2)–(1.8), if any, are required for quadrature formulas on $(0, \infty)$ or (∞, ∞)?

Exercise 3. Prove that the formula analogous to (1.5) for the error using an

$[L/M]$ Padé approximant of a Stieltjes function $f(z)$, as in (1.3), is

$$f(z) - [L/M]_f(z) = \frac{(-z)^{L+M+1}}{\left\{ \tilde{Q}^{[M-1/M]}(z) \right\}^2} \int_a^b \frac{\left\{ \tilde{\pi}_M(u) \right\}^2 u^{J+1} d\phi(u)}{1+uz},$$

where $J = L - M \geqslant -1$, $\tilde{Q}^{[M-1/M]}(z)$ is the denominator of the $[M/M-1]$ Padé approximant of the series

$$g(z) = (-z)^{-J-1} \left[f(z) - \sum_{j=0}^{J} c_j z^j \right] = \int_a^b \frac{u^{J+1} d\phi(u)}{1+zu},$$

and

$$\tilde{\pi}_M(u) = \tilde{Q}^{[M-1/M]}(-1/u)$$

[Chui et al., 1975].

3.2 Tchebycheff's Inequalities for the Density Function

We resume the theme of Part I, Section 5.5, where only the first $k+1$ moments $\mu_0, \mu_1, \ldots, \mu_k$ are given, and these quantities are known to be moments of an unknown Stieltjes density function $\phi(u)$, which means that

$$\mu_j = \int_{-\infty}^{\infty} u^j d\phi(u), \qquad j = 0, 1, 2, \ldots. \tag{2.1}$$

The information contained in the first $k+1$ moments in (2.1) is sufficient to bound the density function. This function $\phi(u)$ is an increasing and piecewise continuous function, conventionally normalized by $\phi(-\infty) = 0$. If $u = u_0$ is a point of discontinuity, then the magnitude of the discontinuity is given by

$$\phi(u_0 +) - \phi(u_0 -) = \Delta\phi(u_0).$$

For a Stieltjes density, $\Delta\phi(u_0) \geqslant 0$, and there is an additive contribution of $\Delta\phi(u_0)\delta(u-u_0)$ in $\phi'(u)$.

The purpose of this section is to derive a piecewise continuous density $\psi(u)$ associated with an arbitrary real point w. It will turn out that w is a point of increase of $\psi(u)$, $\psi(u)$ is nondecreasing and piecewise continuous, and the formula

$$\psi(w-) \leqslant \phi(w) \leqslant \psi(w+) \tag{2.2}$$

provides bounds for $\phi(u)$ at $u = w$.

The general method of approach to this type of problem, begun in Part I, Section 5.5, is to define coefficients

$$c_j = (-)^j \mu_j = (-)^j f_j, \qquad j = 0, 1, 2, \ldots,$$

and a function

$$f(z) = \int_{-\infty}^{\infty} \frac{d\phi(u)}{1 + zu} \tag{2.3}$$

with the formal expansion $f(z) = \sum_{j=0}^{\infty} c_j z^j = \sum_{j=0}^{\infty} f_j(-z)^j$. The next step is the formation of Padé approximants of $f(z)$, from the known coefficients of its expansion. We will need

$$[M/M+1] = \frac{A^{[M/M+1]}(z)}{B^{[M/M+1]}(z)} = \sum_{i=1}^{M+1} \frac{\lambda_i}{1 + zu_i}. \tag{2.4}$$

This Padé approximant is associated with the distribution given by

$$d\psi_{M+1}(u) = \sum_{i=1}^{M+1} \lambda_i \delta(u - u_i), \tag{2.5a}$$

corresponding to a density function

$$\psi_{M+1}(u) = \sum_{i=1}^{M+1} \lambda_i \theta(u - u_i), \tag{2.5b}$$

because these justify the equality

$$[M/M+1] = \int_{-\infty}^{\infty} \frac{d\psi_{M+1}(u)}{1 + zu} = \int_{-\infty}^{\infty} \frac{d\phi(u)}{1 + zu} + O(z^{2M+2}).$$

Had we chosen the $[M/M]$ Padé approximant, we would find

$$[M/M] = \frac{A^{[M/M]}(z)}{B^{[M/M]}(z)} = \lambda_0' + \sum_{i=1}^{M} \frac{\lambda_i'}{1 + zu_i'}, \tag{2.6}$$

which is associated with the distribution given by

$$d\tilde{\psi}_{M+1}(u) = \lambda_0' \delta(u) + \sum_{i=1}^{M} \lambda_i' \delta(u - u_i'), \tag{2.7a}$$

corresponding to the density function

$$\tilde{\psi}_{M+1}(u) = \lambda'_0 \theta(u) + \sum_{i=1}^{M} \lambda'_i \theta(u - u'_i). \tag{2.7b}$$

Notice that these expressions (2.5), (2.7) for $\psi_{M+1}(u)$ and $\tilde{\psi}_{M+1}(u)$ are approximations to $\phi(u)$.

To construct the bounding distribution (2.2), first construct the polynomials

$$g(z) = G(z, w) = B^{[M/M]}(z)B^{[M/M+1]}(w) - B^{[M/M]}(w)B^{[M/M+1]}(z), \tag{2.8}$$

$$h(z) = H(z, w) = A^{[M/M]}(z)B^{[M/M+1]}(w) - B^{[M/M]}(w)A^{[M/M+1]}(z). \tag{2.9}$$

$g(z) = G(z, w)$ is related to the Christoffel–Darboux kernel, and vanishes at $z = w$. Its companion $h(z)$ is chosen so that

$$f(z)g(z) - h(z) = O(z^{2M+1}), \tag{2.10}$$

which follows from the Padé equations. $g(z)$ and $h(z)$ are polynomials of order $M+1$ in z, dependent on the parameter w. Thus we may write

$$\frac{h(z)}{g(z)} = \sum_{j=1}^{M+1} \frac{\rho_j}{1 + z\zeta_j}, \tag{2.11}$$

which is a w-dependent expression, matching $f(z)$ and $M+1$ of its derivatives at the origin. Notice that $\zeta_k = -w^{-1}$ for some k, $1 \leqslant k \leqslant M+1$.

LEMMA 1. *Equation (2.11) defines an $(M+1)$-point quadrature formula which integrates polynomials of order $2M$ exactly.*

Proof. From (2.10) and (2.11),

$$\int_{-\infty}^{\infty} \frac{d\phi(u)}{1 + zu} = \sum_{j=1}^{M+1} \frac{\rho_j}{1 + z\zeta_j} + O(z^{2M+1}) \tag{2.12}$$

Equation (2.12) defines a weighted quadrature formula with points $\zeta_1, \zeta_2, \ldots, \zeta_{M+1}$ and weights $\rho_1, \rho_2, \ldots, \rho_{M+1}$. The accuracy-through-order conditions contained in (2.12) are

$$\int_{-\infty}^{\infty} u^i\, d\phi(u) = \sum_{j=1}^{M+1} \rho_j(\zeta_j)^i, \qquad i = 1, 2, \ldots, 2M,$$

and so polynomials of order no greater than $2M$ are integrated exactly over the measure $d\phi(u)$.

LEMMA 2. *The zeros of $B^{[M/M]}(z)$ and $B^{[M/M+1]}(z)$ interlace.*

Proof. Using (2.4), (2.6) and the ($\ast$) identity,

$$\sum_{i=1}^{M+1} \frac{\lambda_i}{1+zu_i} - \sum_{i=1}^{M} \frac{\lambda_i'}{1+zu_i'} - \lambda_0' = \frac{C(M+1/M)}{C(M/M)} \frac{z^{2M+1}}{B^{[M/M+1]}(z)B^{[M/M]}(z)}$$

$$(2.13)$$

Because $f(z)$ is a Hamburger series,

$$\frac{C(M+1/M+1)}{C(M/M)} z^{2M+1} \lessgtr 0 \qquad \text{for} \quad z \gtrless 0,$$

and the zeros of $B^{[M/M+1]}(z)$ and $B^{[M/M]}(z)$ are real and simple. Let $\{z_i, i=1,2,\ldots, M+1\}$ be the zeros of $B^{[M/M+1]}(z)$, and $\{z_i', i=1,2,\ldots, M\}$ be the zeros of $B^{[M/M]}(z)$. Because $B^{[M/M+1]}(z)B^{[M/M]}(z)$ is a polynomial, the residues of $\{B^{[M/M+1]}(z)B^{[M/M]}(z)\}^{-1}$ alternate in sign. Because $f(z)$ is a Hamburger, λ_i, λ_i' in (2.13) are positive. The interlacing property is a direct consequence of these two observations.

LEMMA 3. *$g(z)$ has real roots.*

Proof. At a root $z = z_i$ of $B^{[M/M+1]}(z)$,

$$\text{sign}\{g(z_i)\} = \text{sign}\{B^{[M/M+1]}(w)\}\, \text{sign}\{B^{[M/M]}(z_i)\}.$$

Thus $g(z)$ has $M+1$ sign changes, and therefore M real roots. Consequently the remaining root is real.

THEOREM 3.2.1. *With the definitions (2.8), (2.9) and (2.11), let*

$$\frac{h(z)}{g(z)} = \frac{H(z,w)}{G(z,w)} = \sum_{i=1}^{M+1} \frac{\rho_i}{1+z\zeta_i}. \qquad (2.14)$$

The lemmas show that (2.14) defines a w-dependent distribution

$$d\psi(u)= \sum_{i=1}^{M+1} \rho_i\delta(u-\zeta_i) \tag{2.15}$$

and a density function

$$\psi(u)= \sum_{i=1}^{M+1} \rho_i\theta(u-\zeta_i). \tag{2.16}$$

$\psi(u)$ *has a point of increase at* $u=\zeta_k = -w^{-1}$, *and*

$$\psi(\zeta_k+)\geqslant\phi(\zeta_k+) \quad and \quad \psi(\zeta_k-)\leqslant\phi(\zeta_k-). \tag{2.17}$$

Proof. Define a polynomial $r^{(k)}(u)$ by

$$r^{(k)}(\zeta_j)=1, \qquad 1\leqslant j\leqslant k \tag{2.18a}$$

$$r^{(k)}(\zeta_j)=0, \qquad k<j\leqslant M+1 \tag{2.18b}$$

$$\frac{d}{du}r^{(k)}(u)\bigg|_{u=\zeta_j} =0, \qquad 1\leqslant j<k \text{ and } k<j\leqslant M+1 \tag{2.19}$$

where the zeros of $g(z)$ have been ordered so that

$$\zeta_1<\zeta_2<\zeta_3< \cdots <\zeta_{M+1}.$$

The polynomial $r^{(k)}(u)$ may be chosen to satisfy the conditions (2.18) and (2.19) with degree at most $2M$. Because $d\psi(u)$ has $M+1$ points of increase,

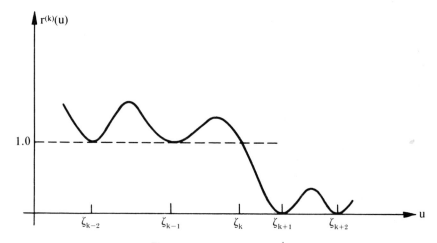

Figure 1. The polynomial $r^k(u)$.

using (2.15), (2.18), we find

$$\psi(\zeta_k +) = \int_{-\infty}^{\infty} r^{(k)}(u)\, d\psi(u).$$

Because the quadrature formula has precision $2M$,

$$\int_{-\infty}^{\infty} r^{(k)}(u)\, d\psi(u) = \int_{-\infty}^{\infty} r^{(k)}(u)\, d\phi(u).$$

Because $\phi'(u) > 0$ and $r^{(k)}(u) \geq 0$,

$$\int_{-\infty}^{\infty} r^{(k)}(u)\, d\phi(u) \geq \int_{-\infty}^{\zeta_k +} r^{(k)}(u)\, d\phi(u).$$

Because $r^{(k)}(u) \geq 1$ on $-\infty < u < \zeta_k +$,

$$\int_{-\infty}^{\zeta_k +} r^{(k)}(u)\, d\phi(u) \geq \int_{-\infty}^{\zeta_k +} d\phi(u) = \phi(\zeta_k +).$$

We conclude that $\psi(\zeta_k +) \geq \phi(\zeta_k +)$. By the use of a different polynomial $t^{(k)}(u)$ for which

$$t^{(k)}(\zeta_j) = 0, \qquad 1 \leq j < k,$$

$$t^{(k)}(\zeta_j) = 1, \qquad k \leq j \leq M+1,$$

the other result that $\psi(\zeta_k -) \leq \phi(\zeta_k -)$ follows directly, and the theorem is proved.

There are manifold applications for these techniques in physics and chemistry, where densities are apt to be positive.

As a postscript to this section, we mention first the Markov problem. Given only the first $k+1$ moments $\mu_0, \mu_1, \ldots, \mu_k$ of an unknown Stieltjes density function $\phi(u)$, so that in principle all the moments are finite and given by

$$\mu_j = \int_{-1}^{1} u^j\, d\phi(u), \qquad j = 0, 1, 2, \ldots,$$

$\phi(u)$ is restricted but not determined by these data. For a given function $h(u)$ and an x satisfying $-1 < x < 1$, the Markov problem is the determination of

$$\inf_{\phi} \int_{-1}^{x} h(u)\, d\phi(u) \quad \text{and} \quad \sup_{\phi} \int_{-1}^{x} h(u)\, d\phi(u).$$

We refer to Shohat and Tamarkin [1963, Chapter 3] for the solution of this problem for various conditions on $h(u)$. The method of solution uses techniques similar to those of this section.

Another interesting problem arises in the theory of equilibrium classical or quantum statistical mechanics, in which systems are described by a distribution over energy E of the form $\exp(-\beta E)\,d\psi(E)$. We are interested in the case where $\beta = 1/kT$, k is the Boltzmann constant, T is the temperature, and so $\beta > 0$. The problem consists of bounding the Stieltjes integral

$$Q(\beta) = \int_0^\infty e^{-\beta E}\,d\psi(E),$$

given the values of the first $2M$ moments of the distribution

$$\mu_i = \int_0^\infty E^i\,d\psi(E), \qquad i = 0, 1, 2, \ldots, 2M-1.$$

This problem was solved by Gordon [1968] using the techniques of this section, but the resulting bounds are not extensive in the system size, and so are only useful for finite-sized systems. Another solution is possible using the methods of Section 1.2. As historical references, we cite Tchebycheff [1874], Stieltjes [1884], and Markov [1884]. More recently, these techniques have been applied to bounding thermodynamic quantities by Isenberg [1963], Gordon [1968], and Corcoran and Langhoff [1977].

3.3 Collocation and the τ-Method

The exponential function, $\exp(x)$, satisfies the differential equation

$$\frac{dy}{dx} - y = 0 \tag{3.1}$$

with the boundary condition

$$y(0) = 1. \tag{3.2}$$

The series solution is $y(x) = \sum_{k=0}^\infty c_k x^k$, and c_k are determined by

$$(k+1)c_{k+1} - c_k = 0, \qquad k = 0, 1, 2, \ldots, \tag{3.3}$$

from (3.1), and $c_0 = 1$ from (3.2). The conventional Lth-order polynomial approximation to $y(x)$, namely $y(x) = \sum_{k=0}^L c_k x^k$, is an exact solution of a modification of (3.1), namely

$$\frac{dy}{dx} - y = \tau x^L. \tag{3.4}$$

with the same boundary condition (3.2). Explicit solution reveals that $\tau = 1/L!$. To obtain a more accurate polynomial approximation to $y(x)$ for use at $x=z$, (3.1) and (3.4) are replaced by

$$\frac{dy}{dx} - y = \tau\pi(x), \tag{3.5}$$

where $\pi(x)$ is an Lth-order polynomial and $\tau\pi(x)$ is small in some sense. Specifically, the τ-method of Lanczos [1957, p. 438] requires that

$$\pi(x) = T_M^*\left(\frac{x}{z}\right) = T_M\left(\frac{2x}{z} - 1\right), \tag{3.6}$$

which is a shifted Tchebycheff polynomial with the famous minimax property (see Part I, Section 6.6) in the range $0 \leqslant x \leqslant z$. We have assumed here that x, z are real and positive, because the analysis is simpler. If z is to be negative or complex, the range in question is $x = \alpha z$, where $0 \leqslant \alpha \leqslant 1$. The introduction of the τ-term into the right-hand side of (3.1) allows solutions which are polynomials in x to be found which are expected to be accurate over the range $0 \leqslant x \leqslant z$.

If $\pi(x)$ is a shifted Legendre polynomial,

$$\pi(x) = P_M^*\left(\frac{x}{z}\right) = P_M\left(\frac{2x}{z} - 1\right), \tag{3.7}$$

then the polynomial solution of (3.5) at the particular point $x=z$ turns out to be the $[M/M]$ Padé approximant of the precise solution of (3.1) at $x=z$ [Luke, 1958, 1960]. We will justify (3.7) before proving the latter result.

One motive for choosing (3.7) in preference to (3.6) is that the explicit algebraic solution is easier to find. We know (Part I, Section 6.6) that the Tchebycheff polynomials $T_m(x)$ satisfy the minimax property for the class $\mathcal{P}_n$ of polynomials of order n, $n \geqslant 1$, with leading coefficient unity:

$$\inf_{p_n(x) \in \mathcal{P}_n} \sup_{-1 \leqslant x \leqslant 1} |p_n(x)| = 2^{-(n-1)} \tag{3.8}$$

is achieved when $p_n(x) = T_n(x)$. In this sense, the Legendre polynomials are remarkably small. With their conventional normalization

$$P_n(x) = \frac{1}{2^n n!}\left(\frac{d}{dx}\right)^n (x^2 - 1)^n, \tag{3.9}$$

it is well known that [Isaacson and Keller, 1966, p. 219]

$$\sup_{-1 \leqslant x \leqslant 1} |P_n(x)| = 1 \tag{3.10}$$

and the leading coefficient is derived from (3.9) as

$$\frac{(2m)!}{2^m(m!)^2} = \frac{2^m}{\sqrt{m\pi}}\left[1+O(m^{-1})\right]. \tag{3.11}$$

Equations (3.9), (3.10), and (3.11) may be "renormalized" to show comparability with (3.8). The Legendre polynomials have all their zeros in the interior of $[-1,1]$; furthermore, the optimal polynomial for the right-hand side of (3.5) is unlikely to be independent of the specific structure of the equation, and thus selection of $\pi(x)=P_M^*(x/z)$ is to be regarded as a very satisfactory choice.

To be a little more general than choosing $\pi(x)=P_M^*(x/z)$, we choose

$$\pi(x)=x^J G_M(J+1, J+1, x/z). \tag{3.12}$$

For brevity we write $G_M(J+1, J+1, x/z)=R_M^J(x/z)$, which is a Jacobi polynomial normalized to have leading coefficient unity. The Jacobi polynomials $G_M(p,q;x)$ are orthogonal on $(0,1)$ over a measure $w(x)=(1-x)^{p-q}x^{q-1}$, and so our choice of $R_M^J(x/z)$ selects polynomials orthogonal on $(0,z)$ over the measure x^J.

The special case of $J=0$, with $R_M^0(x/z)\propto P_M^*(x/z)$, is of special interest, since it corresponds to (3.7).

From Abramowitz and Stegun [1964, p. 775], we find that

$$R_M^J\left(\frac{x}{z}\right)=\frac{(J+M)!}{(J+2M)!}\sum_{m=0}^{M}(-)^m\,{}^MC_m\frac{(J+2M-m)!}{(J+M-m)!}\left(\frac{x}{z}\right)^{M-m} \tag{3.13}$$

$$=\sum_{k=0}^{M}c_k^{(M,J)}\left(\frac{x}{z}\right)^k$$

and we must solve (3.5) and (3.12) as

$$\frac{dy}{dx}-y=\tau x^J R_M^J\left(\frac{x}{z}\right) \tag{3.14}$$

for a polynomial solution $y(x)=\sum_{k=0}^{L}d_k x^k$, subject to consistency. Equating coefficients of x^k on each side of (3.14), we find

$$(k+1)d_{k+1}-d_k=\tau c_{k-J}^{(M,J)}z^{-k+J} \quad \text{for} \quad k=J, J+1,\ldots, L, \tag{3.15a}$$

$$(k+1)d_{k+1}-d_k=0 \quad \text{for} \quad k=0,1,\ldots, J-1. \tag{3.15b}$$

From (3.15b),

$$d_k=\frac{J!}{k!}d_J \quad \text{for} \quad k=0,1,\ldots, J-1, \tag{3.16a}$$

and from (3.15a), because $y(x)$ is a polynomial and $d_{L+1}=0$, it follows that

$$d_k = -\tau \sum_{r=k-J}^{M} c_r^{(M,J)} \frac{(r+J)!}{k!} z^{-r} \quad \text{for} \quad k=J, J+1,\ldots, L.$$

$$(3.16b)$$

Substitute for $c_r^{(M,J)}$ from (3.13), and then (3.16) shows that

$$d_0 = -\tau z^{-M} L! \, {}_1F_1(-M, -(L+M); -z).$$ (3.17)

Since $d_0 = 1$ is the boundary condition (3.2), τ is determined and is given by

$$\tau = -z^M [L! \, {}_1F_1(-M, -(L+M); -z)]^{-1}.$$ (3.18)

At this stage it is clear that the solution of (3.14) is a polynomial in x at the point $x=z$, the solution is a rational function of z, and the degree of its denominator is M.

Calculation of the full solution proceeds by using (3.16b) to prove

$$d_J = -\frac{\tau z^{-M} L!}{J!(L+M)!} \sum_{j=0}^{M} (-)^j \frac{M!(L+M-j)!}{(M-j)!j!} z^j.$$ (3.19)

It then follows from (3.16) and (3.19) that

$$y(z) = -\frac{\tau z^{-M} M! L!}{(L+M)!} \sum_{j=0}^{M} \sum_{k=0}^{L-j} \frac{z^{j+k}}{k!} \frac{(-)^j(L+M-j)!}{(M-j)!j!}$$

$$= (-\tau) z^{-M} L! \, {}_1F_1(-L, -(L+M); z).$$ (3.20)

Eliminating τ from (3.18) and (3.20) yields

$$y(z) = \frac{{}_1F_1(-L, -(L+M); z)}{{}_1F_1(-M, -(L+M); -z)},$$ (3.21)

which is an $[L/M]$-type rational function of z. We can show that it is, in fact, the $[L/M]$ Padé approximant of $y(z)$, either by referring to (I.1.2.12) or by proving that

$$y(z) - \exp(z) = O(z^{L+M+1})$$ (3.22)

by utilizing the present derivation.

Equation (3.14) is simplified, using an integrating factor, as

$$\frac{d}{dx}(e^{-x}y)=e^{-x}\tau x^J R_M^J\left(\frac{x}{z}\right).\tag{3.23}$$

Define $\varepsilon(x)=y(x)-\exp(x)$; since

$$\frac{d}{dx}(e^{-x}\exp(x))=0,$$

integration of (3.23) yields

$$\varepsilon(z)=e^z\int_0^z e^{-x}\tau x^J R_M^J\left(\frac{x}{z}\right)dx$$

$$=\tau e^z z^{J+1}\int_0^1 e^{-zt}R_M^J(t)t^J\,dt\tag{3.24}$$

with the boundary condition $\varepsilon(0)=0$ built in.

Using Rodrigues's formula (3.9), it follows that

$$\int_0^1 e^{-zt}R_M^J(t)t^J\,dt=O(z^M),\tag{3.25}$$

and hence $\varepsilon(z)=O(\tau z^{L+1})=O(z^{L+M+1})$, verifying (3.22).

In summary, (3.14) with $J=L-M$ defines an $[L/M]$ Padé approximant for $\exp(z)$ for $L\geqslant M$, and the explicit solution is given by (3.21). Apparently, the optimal solution is given by $J=0$. The choice $J=L$, $M=0$ gives the ordinary Maclaurin polynomial solution.

Let us consider exactly what has been achieved in finding an approximate solution of (3.1) by the method given, which was introduced as the Lanczos τ-method, as described earlier.

The method is also a collocation method [Wright, 1970]. Points $\{x_i, i=1,2,\ldots,L\}$, not necessarily distinct, which are in fact the zeros of $\pi(x)$ are selected in the interval $[0, z]$. The approximate solution is the polynomial solution $y_L(x)$ of order L which satisfies

$$\frac{dy_L}{dx}-y_L=0\qquad\text{at}\quad x=x_i,\ i=1,2,\ldots,L.\tag{3.26}$$

These hypotheses imply directly that

$$\frac{dy_L}{dx}-y_L=\tau\pi(x),$$

returning us to our starting point (3.5), and showing that the method is a collocation method. Implementation of the collocation method consists primarily of determining the coefficients d_j of the monomials x^j which constitute the polynomial $y_L(x) = \Sigma_{j=0}^{L} d_j x^j$.

The method may be viewed as an implicit Runge–Kutta method, using Gauss points $\{x_i\}$ in the interval $[0, z]$. The method returns a quadrature solution of

$$y(z) = 1 + \int_0^z y(x)\, dx$$

taking the form, with $x_j = z\xi_j$,

$$\sum_{r=0}^{L} c_r (z\xi_j)^r = 1 + \sum_{r=0}^{L} \frac{c_r (z\xi_j)^{r+1}}{r+1}, \qquad j = 1, 2, \ldots, n.$$

It follows that we have L implicit equations for L unknown coefficients $\{c_r, r = 1, \ldots, n\}$. The function and its derivative are implicitly determined pointwise at the nodes.

In summary, the collocation method and the implicit Runge–Kutta method are equivalent in the present context; either leads to a solution which is a rational function of z. If this solution can be shown to satisfy the accuracy-through-order condition (3.22), and if z is identified as the step-length (commonly called h), then the solution is a Padé approximant. Furthermore, these methods are equivalent to certain Galerkin methods in which the inner products are evaluated by Gauss–Legendre quadrature: this is the collocation method in a different guise. The connection between implicit methods and Padé approximation appears again in the next section.

If we apply the method to the hypergeometric equation,

$$\left[(x - x^2) \frac{d^2}{dx^2} + \gamma \frac{d}{dx} - x(1 + \alpha + \beta) \frac{d}{dx} - \alpha\beta \right] y = 0, \qquad (3.27)$$

we encounter immediately the problem that (3.27) is a second-order equation. But by writing it as

$$\frac{d}{dx} \left[(x - x^2) \frac{dy}{dx} + \{\gamma - 1 - x(\alpha + \beta - 1)\} y \right] + (\alpha + \beta - 1 - \alpha\beta) y = 0,$$

we see that it can be reduced to a first-order differential equation if $(\alpha - 1)(\beta - 1) = 0$. We take $\beta = 1$, and consider the solution of

$$(x - x^2) \frac{dy}{dx} + (\gamma - 1 - \alpha x) y = (\gamma - 1) y(0). \qquad (3.28)$$

The τ-method is implemented by choosing

$$(1-x)\frac{dy}{dx}+\left(\frac{\gamma-1}{x}-\alpha\right)y=\tau P\left(\frac{x}{z}\right)x^J+\frac{(\gamma-1)y(0)}{x}, \qquad (3.29)$$

where $P(x)$ is a polynomial of order M to be specified. Equation (3.29) corresponds to modifying (3.27) to

$$(x-x^2)\frac{d^2y}{dx^2}+\gamma\frac{dy}{dx}-x(2+\alpha)\frac{dy}{dx}-\alpha y=\tau\frac{d}{dx}\left\{x^{J+1}P\left(\frac{x}{z}\right)\right\},$$

but this is not directly useful.

To integrate (3.29), rewrite it as

$$\frac{dy}{dx}+\left[(\gamma-1)\left(\frac{1}{x}+\frac{1}{1-x}\right)-\frac{\alpha}{1-x}\right]y=\frac{\tau x^J}{1-x}P\left(\frac{x}{z}\right)+\frac{(\gamma-1)y(0)}{x(1-x)}.$$

$$(3.30)$$

The integrating factor is $x^{\gamma-1}(1-x)^{1-\gamma+\alpha}$, and so (3.30) becomes

$$\frac{d}{dx}\left\{x^{\gamma-1}(1-x)^{1-\gamma+\alpha}y\right\}=\tau x^{J+\gamma-1}(1-x)^{\alpha-\gamma}P\left(\frac{x}{z}\right)$$

$$+(\gamma-1)x^{\gamma-z}(1-x)^{\alpha-\gamma}y(0). \qquad (3.31)$$

Integrating (3.31), and checking consistency at $x=0$, leads to

$$y(x)=x^{1-\gamma}(1-x)^{\gamma-1-\alpha}$$

$$\times\int_0^x\left\{\tau t^{J+\gamma-1}(1-t)^{\alpha-\gamma}P\left(\frac{t}{z}\right)+(\gamma-1)t^{\gamma-z}(1-t)^{\alpha-\gamma}y(0)\right\}dt$$

$$(3.32)$$

Naturally, $F(x)={}_2F_1(\alpha,1,\gamma;x)$ satisfies a similar integral equation with $\tau=0$, and hence $\varepsilon(x)=y(x)-F(x)$ obeys

$$\varepsilon(x)=x^{1-\gamma}(1-x)^{\gamma-1-\alpha}\int_0^x\tau t^{J+\gamma-1}(1-t)^{\alpha-\gamma}P\left(\frac{t}{z}\right)dt$$

Hence

$$\varepsilon(z)=z^{J+1}\tau(1-z)^{\gamma-1-\alpha}\int_0^1 x^{J+\gamma-1}(1-xz)^{\alpha-\gamma}P(x)\,dx, \qquad (3.33)$$

and the correct choice of $P(x)$ is that of a Jacobi polynomial orthogonal over weight $x^{J+\gamma-1}$, and we assume that $J+\gamma>0$. From Abramowitz and Stegun [1964, p. 774] we find

$$P(x)=\frac{\Gamma(\gamma+J)}{\Gamma(\gamma+J+M)}\sum_{m=0}^{M}(-)^{m}{}^{M}C_{m}\frac{\Gamma(\gamma+J+2M-m)}{\Gamma(\gamma+J+M-m)}t^{M-m},$$

which reduces to (3.13) when $\gamma=1$. An analysis almost identical to that following (3.13) reveals that

$$\tau^{-1}=z^{-M}{}_{2}F_{1}\big(-(\alpha+L),-M,-(\gamma+L+M-1);z\big),\qquad(3.34)$$

and we deduce that

$$Q^{[L/M]}(z)={}_{2}F_{1}\big(-(\alpha+L),-M,-(\gamma+L+M-1);z\big).\qquad(3.35)$$

The numerator $P^{[L/M]}(z)$ does not have a simple compact form in general. That these formulas are the numerators and denominators of Padé approximants follows from (3.33) and (3.34), since $\varepsilon(z)=O(z^{J+1+M+M})=O(z^{L+M+1})$. It is interesting to compare this derivation of (3.35), the Padé denominator of ${}_{2}F_{1}(\alpha,1;\gamma;z)$, with the methods of Part I, Section 1.2, and the continued-fraction method based on (I.4.6.14).

For a complete review of the explicit formulas obtainable using the techniques of this section, we refer to Luke [1964, 1969]. For further details of the connection between the τ-method, collocation, and implicit Runge–Kutta methods, we refer to Butcher [1963, 1964], Ehle [1968], Mason [1967], Axelsson [1969, 1972], and Chipman [1971].

3.4 Crank–Nicholson and Related Methods for the Diffusion Equation

The initial-value problem associated with the diffusion (or heat) equation is the problem of finding a continuous function $u(x,t)$ which satisfies

$$\frac{1}{\kappa}\frac{\partial u}{\partial t}-\frac{\partial^{2}u}{\partial x^{2}}=0,\qquad t>0,\qquad(4.1a)$$

subject to the boundary condition

$$u(x,0)=f(x),\qquad-\infty<x<\infty.\qquad(4.1b)$$

These equations define a mathematical problem (a pure initial-value problem), which describes many idealized diffusion processes, among them being the process of heat transfer along a bar. The temperature $u(x,0)=f(x)$ is specified at the initial time $t=0$, and the problem is to find the temperature at later times $t>0$.

From Green's function associated with (4.1), the solution to (4.1) is well known to be

$$u(x,t) = \frac{1}{\sqrt{4\pi\kappa t}} \int_{-\infty}^{\infty} \exp\left\{ \frac{-(\xi-x)^2}{4\kappa t} \right\} f(\xi)\, d\xi.$$

Thus the solution of (4.1) is reduced to numerical integration for all times $t>0$.

There is, of course, considerable interest in solving (4.1a) by other methods which can be used when the Green's function method is inapplicable, for instance when the equation is slightly different or the boundary conditions are different. From reactor physics, we might wish to consider the following partial differential equation for the neutron density $u(\mathbf{x},t)$:

$$\underset{\text{time variation}}{\frac{1}{\kappa}\frac{\partial u}{\partial t}} - \underset{\text{diffusion}}{\nabla^2 u} = \underset{\text{source term}}{s(\mathbf{x},t)u}.\qquad(4.2)$$

We might consider heat transfer along a finite bar, with the ends maintained at given temperatures and with various heat sources; the temperature $u(x,t)$ obeys an equation such as

$$\frac{1}{\kappa}\frac{\partial u}{\partial t} - \frac{\partial^2 u}{\partial x^2} = s(x,t), \qquad 0<x<1,\qquad(4.3a)$$

with the boundary conditions at the ends of the bar

$$\begin{aligned} u(0,t) &= T_0(t), \\ u(L,t) &= T_L(t), \end{aligned}\qquad(4.3b)$$

and the specification of the initial temperature

$$u(x,0)=f(x).$$

It is obvious enough from the physics of the heat problem that (4.3b) and (4.3c) are sufficient boundary conditions for the solution of (4.3a). The boundary conditions necessary with a mathematical formulation are not always quite so obvious.

To obtain numerical solutions of equations such as (4.2) and (4.3) and others of that type, various methods are available, e.g. Galerkin methods, but the simplest is to replace the derivatives by differences, and initially we discretize the x-variable in (4.3). We choose N interior points in $0<x<L$,

so that the mesh spacing Δx is given by $\Delta x = L/(N+1)$ and the points used are given by

$$x_i = i\Delta x, \qquad i = 1, 2, \ldots, N.$$

The approximation scheme (method of lines) is defined by

$$u(x_i, t) \to U_i(t)$$

and

$$\left.\frac{\partial^2 u}{\partial x^2}\right|_{x=x_i} \to \frac{U_{i+1}(t) - 2U_i(t) + U_{i-1}(t)}{(\Delta x)^2}.$$

We solve the "space discretized" equations

$$\frac{1}{\kappa}\frac{\partial U_i}{\partial t} - \frac{U_{i+1} - 2U_i + U_{i-1}}{(\Delta x)^2} = s(x_i, t) \tag{4.4}$$

for the functions $U_1(t), U_2(t), \ldots, U_N(t)$. It is understood that $U_0(t) = T_0(t)$ and $U_{N+1}(t) = T_L(t)$. It is important to be able to prove that this system of equations is convergent; this statement means that as $\Delta x \to 0$ and $N \to \infty$, the approximation scheme implemented with exact arithmetic tends to the exact result. However, we will not attempt this proof here.

Equation (4.4) can be written in matrix form:

$$\frac{\partial U_i}{\partial t} = \sum_{j=1}^{N} A_{ij} U_j(t) + S_i(t), \qquad i = 1, 2, \ldots, N, \tag{4.5}$$

where $S_i(t) = \kappa S(x_i, t) + \kappa\{\delta_{i1}U_0(t) + \delta_{iN}U_{N+1}(t)\}/(\Delta x)^2$.

If we had been treating higher-dimensional systems, such as the three-dimensional reactor problem of (4.2), we would have been led to an equation similar to (4.5). However, the source terms would be different, and, more significantly, the matrix A would be a large sparse matrix. A sparse matrix has most of its entries zero, which means that special techniques may be used to avoid redundant arithmetic operations. The practical problem we have in mind is that of treating the case where A is a not necessarily symmetric large matrix of dimension $N^3 \times N^3$.

Equation (4.5) typifies diffusion equations after space discretization, and the all too obvious method of solution is by time discretization. We use a sequence of time points $t_0 = 0, t_1, t_2, t_3, \ldots$ and let $\Delta t_k = t_{k+1} - t_k$. The approximations consist of

$$U_i(t_k) \rightarrow U_i^{(k)}, \qquad k=1,2,\ldots,$$

$$\left.\frac{\partial U_i}{\partial t}\right|_{t=t_k} \rightarrow \frac{U_i^{(k+1)} - U_i^{(k)}}{\Delta t_k}.$$

Then (4.5) is replaced by the equation

$$U_i^{(k+1)} = U_i^{(k)} + \Delta t_k \left\{ \sum_{j=1}^{N} A_{ij} U_j^{(k)} + S_i(t_k) \right\}, \tag{4.6}$$

which is represented schematically in Figure 1.

Let us now consider the form of the equations with all boundary and source terms set to zero. The major problems we will discover and solve are independent of these terms, and their presence would be largely immaterial. Then (4.6) is temporarily replaced by

$$\mathbf{U}^{(k+1)} = (I + \Delta t \, A)\mathbf{U}^{(k)}, \tag{4.7}$$

where $\mathbf{U}^{(k)} = (U_1^{(k)}, U_2^{(k)}, U_3^{(k)}, \ldots, U_N^{(k)})$ is a vector of values at time t_k. To discover the effect of the matrix operator $(I + \Delta t \, A)$, let us take the case of $N=3$; using (4.4) we write

$$I + \Delta t \, A = I - \frac{\kappa \Delta t}{(\Delta x)^2} B_3,$$

where

$$B_3 = \begin{pmatrix} 2 & -1 & 0 \\ -1 & 2 & -1 \\ 0 & -1 & 2 \end{pmatrix}.$$

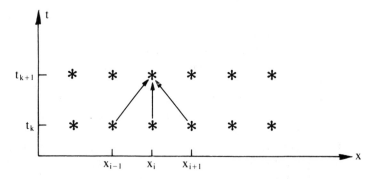

Figure 1. Diagrammatic representation of space and time points relevant to (4.6).

For arbitrary N, we would define

$$(B_N)_{ij} = 2\delta_{ij} - \delta_{i-1,j} - \delta_{i,j-1}.$$

The eigenvalues of the matrix B_3 are easily found to be 2 and $2 \pm \sqrt{2}$. In general, the eigenvalues of B_N are positive and are given by

$$\lambda_i = 4\sin^2\left(\frac{\pi}{2} \cdot \frac{i}{N+1}\right), \qquad i = 1, 2, \dots, N. \qquad (4.8)$$

It follows from Part I, Section 4.4, using a similar recurrence relation to (I.4.4.5) for tridiagonal determinants, that the polynomial given by $\det(B_N - \lambda I)$ is simply related to Nth-order shifted Tchebycheff polynomials, yielding (4.8).

The essential point is that the eigenvalues of $(I + \Delta t\, A)$ are given by

$$\Lambda_i = 1 - \frac{\kappa \Delta t}{(\Delta x)^2} \lambda_i.$$

Since the operator $(I + \Delta t\, A)$ in (4.7) is to be used iteratively, an essential requirement for stability is that $|\Lambda_i| < 1$. Otherwise, rounding errors will grow exponentially during the iteration. This stability condition implies that

$$0 < \frac{\kappa \Delta t}{(\Delta x)^2} \lambda_i < 2, \qquad (4.9)$$

and from this inequality we see that a necessary condition for stability is that the time step is limited by

$$\Delta t < 2(\Delta x)^2 / \kappa. \qquad (4.10)$$

Consequently, the explicit method defined by (4.7) is called conditionally stable. The upper limit on the time step given by (4.10) and imposed by the requirement of stability is unsatisfactory, and we will find a way of avoiding it.

We know, from the physics of the heat problem (4.3), that the initial behavior is transient heat flow in the bar and that then the system "settles down". This suggests that the final time steps may be much larger than the initial ones with a satisfactory method. In fact, the nature of the true solution of later times is dominated by the effect of the smallest eigenvalue λ_1 of (4.8) corresponding to a time variation of $\exp(-\lambda_1 \Delta t)$ between steps. Systems in which the numerical solution is controlled by a large eigenvalue and the true solution is controlled by a much smaller eigenvalue are called *stiff systems*. Clearly, our present explicit method of solution of such systems will not do.

Let us review our intentions to see the solution to the dilemma, by looking at (4.5), with no source terms, as our idealized problem. It is

$$\frac{\partial U_i}{\partial t} = \sum_{j=1}^{N} A_{ij} U_j, \tag{4.11}$$

where A is a known constant matrix with negative eigenvalues. Equation (4.11) has the solution

$$\mathbf{U}^{(k+1)} = \exp(A \Delta t) \mathbf{U}^{(k)}, \tag{4.12}$$

and so the question becomes that of how the matrix $\exp(A \Delta t)$ is to be calculated.

Method 1. The first attempt might be to use a truncated Maclaurin series. The prime objection to this method is that it is unstable. It is also computationally expensive. Matrix multiplication with $N \times N$ matrices requires $O(N^3)$ multiplications, and so $r+1$ terms of the series require $O((r-1)N^3)$ multiplications (ignoring any benefits of sparseness).

Method 2. The second attempt might be formation of the matrix $\{\exp(-A \Delta t)\}^{-1}$ using a truncated Maclaurin expansion, which is stable but just as expensive as the former method. Methods such as this, requiring matrix inversion, are called *implicit methods*, for reasons which will become clear. Remember that inversion requires $O(N^3)$ multiplications and divisions (using Gaussian elimination) and so is not unduly expensive on the relevant scale; direct solution of matrix equations requires $O(\frac{1}{3} N^3)$ operations, and this saving is always important in practice. Also remember that there are special methods with $O(2N)$ operations for tridiagonal matrices occurring in one (and higher) dimensional problems.

Method 3. A possibility is the use of $[1+(1/n)A \Delta t]^n$ with n sufficiently large, which reduces the computational effort, but this is unstable, and the third attempt would be the use of $[1-(1/n)A \Delta t]^{-n}$. Whilst it is true that $[1-(1/n)A \Delta t]^n \rightarrow \exp(A \Delta t)$ as $n \rightarrow \infty$, only the first two terms of the Maclaurin series agree, and so this is a first-order (stable) method. The method is realistically viewed as n-fold application of $(1-A \delta t)^{-1}$ using a smaller time step $\delta t = \Delta t / n$.

Method 1 can be regarded as the method using the $[r/0]$ Padé approximant to $\exp(A \Delta t)$, and method 2 can be regarded as the method using the $[0/r]$ Padé approximant. We view method 3 as the method corresponding to repeated use of the $[0/1]$ Padé approximant.

These considerations motivate investigation of other Padé approximants, and the $[1/1]$ is an obvious candidate. The approximation to (4.12) takes the form

$$\mathbf{U}^{(k+1)} = \left(1 - \tfrac{1}{2} A \, \Delta t\right)^{-1}\left(1 + \tfrac{1}{2} A \, \Delta t\right)\mathbf{U}^{(k)} \tag{4.13}$$

and is called the Crank–Nicholson method. It is a second-order method, being accurate through order $(\Delta t)^2$. If ξ is an eigenvalue of $A\,\Delta t$, then

$$\Xi(\xi) = \frac{1 + \tfrac{1}{2}\xi}{1 - \tfrac{1}{2}\xi} \tag{4.14}$$

is the eigenvalue of $(1 - \tfrac{1}{2} A\,\Delta t)^{-1}(1 + \tfrac{1}{2} A\,\Delta t)$. Because $\xi < 0$, (4.14) shows that $|\Xi(\xi)| < 1$. This feature of the method is most important because it shows that rounding errors are not magnified during the iterative solution. Such a process is A-stable, or absolutely stable, and is an essential prerequisite for a satisfactory numerical solution. A similar calculation for methods 1 and 2 reveals their respective instability and stability.

DEFINITION. Suppose that the values $\{y_n, \, n = 0, 1, 2, \ldots\}$ are obtained as the numerical solution of the test equation $y' = \lambda y$, so that $y_n \approx y(nh)$. The numerical method is *A-stable* if $y_n \to 0$ as $n \to \infty$ for every fixed λ, h with Re $\lambda < 0$ and $h > 0$.

Another desirable feature of a method is L-stability, sometimes called left-stability, stiff-stability, or stiff A-stability to add mystery and confusion.

DEFINITION. Suppose that the values $\{y_n, n = 0, 1, 2, \ldots\}$ are obtained as the numerical solution of the test equation $y' = \lambda y$, so that $y_n \approx y(nh)$ and $y_{n+1} = f(h\lambda)y_n$ with Re $\lambda < 0$ and $h > 0$. The numerical method is *L-stable* if it is A-stable and $f(h\lambda) \to 0$ as Re$(h\lambda) \to -\infty$.

These definitions are phrased in the standard notation for ordinary differential equations [Gear, 1971; Lambert, 1974]. The definition of L stability [Axelsson, 1969] uses the result that a numerical method for the solution of an ordinary differential equation is closely linked to a rational approximation $f(h\lambda)$ of $\exp(h\lambda)$, as we have seen in (4.12), (4.13) for systems of such equations. The example of (4.2)–(4.14) demonstrates that A-stability is equivalent to the requirement that $|f(h\lambda)| < 1$ whenever Re$(h\lambda) < 0$, and that errors do not grow during computation [Lambert, 1974].

Let $[L/M](x)$ be the $[L/M]$ Padé approximant of the exponential function. It is known that $|[L/M](x)| < 1$ for all Re $x < 0$ if $L = M$ [Varga, 1961] or $L = M - 1$ or $L = M - 2$ [Ehle, 1973]. Hence we see that $[M/M]$, $[M-1/M]$ and $[M-2/M]$ Padé methods are A-stable. By power counting, we see that the $[M-1/M]$ and $[M-2/M]$ methods are L-stable. The

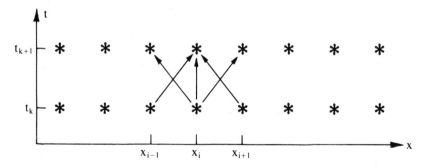

Figure 2. Diagrammatic representation of heat flow at the point x_i at sequential times.

additional requirment that $[L/M](-\infty)=0$ for L-stability may be regarded as a constraint from two point Padé approximation.

In certain cases, as (4.8) shows, one can predetermine the eigenvalues of A. If each eigenvalue λ satisfies $\mathrm{agr}(\lambda)<\alpha<\pi/2$, then the numerical method can be guaranteed stability under the milder restriction of $A(\alpha)$ stability. This corresponds to the condition that $|[L/M](x)|<1$ for $|\arg(-x)|<\alpha$. We refer to Gear [1971] for details.

The Crank–Nicholson method (4.13) was introduced as a [1/1] Padé-approximant method. It might have been introduced by considerations motivated by Figure 2. The arrows show the direction of heat flow causing loss or gain of heat at the point x_i at sequential times. Alternatively, the Crank–Nicholson method may be motivated by asking for a more accurate estimate of $\partial U_i/\partial t$ in (4.11) using differences centered on $t_k+\frac{1}{2}\Delta t_k$ rather than t_k. This facet of the argument may be regarded either as the common-sense approach, or as a second-order $[O((\Delta t)^2)]$ method, or else as the method derived from the trapezoidal rule for the integrated form of (4.9). It is the most accurate A-stable linear implicit multistep method. Even though it is not L-stable, the Crank–Nicholson method is well recommended for many reasons.

Hitherto, we have concentrated primarily on achieving accuracy (maximising the order of the local truncation error) and on stability of the numerical solution of diffusion equations. However, it is as well to recall all the objectives and practicalities of the problem. Our aim is to choose the most satisfactory rational approximation to $\exp(x)$ on $[0, \Delta t]$, and a Padé approximant is only "best" in the limit $\Delta t \to 0$ (cf. Section 3.6). A practical consideration is inexpensive construction of the rational form, and to this end it is desirable that the matrix approximant of $\exp(A\Delta t)$ should have real linear factors. In the full context, Padé methods are relevant as a means of contrasting certain virtues of different schemes. Only experience will decide which are the best methods of resolving the difficulties caused by variegated and conflicting requirements for any particular problem.

We refer to the books of Crank [1975], Carslaw and Jaeger [1959], and Bell and Glasstone [1970] for accounts of the origins of the parabolic equations discussed in this section. We refer to the books of Varga [1962], Gear [1971], and Lambert [1975] and the review of Argyris et al. [1977] for accounts of the methods of solution of the equations. We refer to Baker [1960], Baker and Oliphant [1960], Ehle [1968, 1973, 1976], Cavendish et al. [1972], Fairweather [1971], Nørsett [1974], Siemeniuch [1976], Saff et al. [1976], Smith et al. [1977], and Saff et al. [1977] as a selected set of references on the topics of this section.

Exercise Prove that (4.8) is valid.

3.5 Inversion of the Laplace Transform

The problem of inversion of the Laplace transform is the determination of $f(t)$, given $\bar{f}(p)$. In principle, the functions are related by

$$f(t) = \frac{1}{2\pi i} \int_{\omega - i\infty}^{\omega + i\infty} \bar{f}(p) e^{pt} dp \tag{5.1}$$

and

$$\bar{f}(p) = \int_0^\infty e^{-pt} f(t) dt. \tag{5.2}$$

Equation (5.1) determines $f(t)$ from $\bar{f}(p)$ directly, provided a suitable contour is chosen on which the integrand is not unduly oscillatory. The trouble with this problem is that small oscillations in $\bar{f}(p)$ result from large changes in $f(t)$. Here we discuss only the Padé method, which avoids the task of selecting a suitable contour and involves an implicit smoothness assumption. To motivate the method, we follow the original problem first solved this way, which has smoothness properties built in by the underlying physics.

Waves moving in an imperfectly elastic medium are conjectured to satisfy the equation for the displacement $y(x, t)$

$$\frac{\partial^2 y}{\partial t^2} = c^2 \left(1 + \tau \frac{\partial}{\partial t}\right) \frac{\partial^2 y}{\partial x^2}. \tag{5.3}$$

Under initial conditions representing propagation of a transverse displacement caused by a unit step function displacement at time $t=0$ at the end of the bar, $x=0$, we use the Heaviside function $\theta(t)$ to define

$$y(x=0, t)=y_0(t)=\theta(t).\tag{5.4}$$

Using the transform

$$\bar{y}(x, p)=\int_0^\infty e^{-pt}y(x, t)\,dt,$$

Equation (5.3) is simplified to become

$$\frac{\partial^2\bar{y}(x, p)}{\partial x^2}=\frac{p^2}{c^2(1+\tau p)}\bar{y}(x, p)\qquad\text{for}\quad x>0,$$

with the solution

$$\bar{y}(x, p)=\bar{y}_0(p)\exp\left\{\frac{-px}{c\sqrt{1+\tau p}}\right\}$$

$$=\frac{1}{p}\exp\left\{\frac{-px}{c\sqrt{1+\tau p}}\right\}.\tag{5.5}$$

The solution for $y(x, t)$ requires the inverse transform of (5.5). $\bar{y}(x, p)$ is meromorphic in p for $\operatorname{Re} p>-\tau^{-1}$, and analytic in $\operatorname{Re} p>0$. Hence (5.1) is well defined in this application if $\omega>0$. We omit the changes of variable which are, in practice, desirable for putting (5.5) in canonical form, because they are inessential to the description. The Padé method requires three simple steps.

Step 1.   Write (5.5) as a Maclaurin series

$$p\bar{y}(x, p)=c_0+c_1p+c_2p^2+\cdots,$$

and form its diagonal Padé approximants $[M/M](p)$. The coefficients depend on x and the parameters c, τ.

Step 2.   Reexpress this Padé approximant as

$$\frac{1}{p}[M/M](p)=\sum_{i=0}^M\frac{\gamma_i^{(M)}(x)}{p-p_i^{(M)}(x)},\qquad\text{where}\quad p_0^{(M)}(x)=0.\tag{5.6}$$

If $\operatorname{Re} p_i^{(M')}(x)>0$ for any pair (i, M'), then the $[M'/M']$ Padé approximant is an unacceptable approximation because it has the wrong analytic structure. The poles of the Padé approximants ought to be close to the essential singularity and the natural locations of the cuts of (5.5), which are in the left half plane. Furthermore, a pole in the right half plane would lead to an unphysical solution which increases indefinitely with time.

Step 3. The approximate inverse transform is given by the transform of (5.6), and is

$$y(x,t) \approx \sum_{i=0}^{M} \gamma_i^{(M)}(x) e^{p_i^{(M)}(x)t}.$$

This is a sum of damped exponentials; it is interesting to compare this method with that of the Baker–Gammel approximants (Section 1.2) with an exponential kernel.

We make no claim that the method described in this section is a "best method" of inversion of the Laplace transform. Improvements based on development of this method incorporate a best or near-best rational approximation of $\bar{f}(p)$ in (5.1) [Longman, 1974; Brezinski, 1979]. We refer to Talbot [1979] for a different method and a list of methods which achieve good accuracy in general cases.

3.6 Connection with Rational Approximation

We have occasionally emphasised in earlier sections that Padé approximants are not usually best rational approximants. It is notoriously difficult, in general, to find best rational approximants either in principle or numerically.

For a function $f(z)$ whose singularities are confined to a finite number of regions, Runge's theorem, very crudely stated, shows that rational approximants exist having prescribed poles in the singularity regions only, and which converge to $f(z)$ in the complement regions where $f(z)$ is holomorphic. To explain what is easily attainable using rational approximation, we prove two introductory lemmas, and then we carefully state and prove Runge's theorem. This presentation is followed by a simple example showing the futility of polynomial approximation in a particular case for which rational approximation is exact. Finally we present Walsh's theorem and some associated results showing that Padé approximants are the best local rational approximants.

LEMMA 1. *Let $\mathcal{D}$ be an open, connected, and bounded domain, with a boundary curve C consisting of an outer boundary C_0 and inner boundary curves $C_1, C_2, \ldots, C_n$. Let $f(z)$ be holomorphic in $\mathcal{D}$ and continuous on C. Then a sequence $\{P_k(z)\}$ of rational functions exist whose poles lie on C, and $P_k(z) \to f(z)$ uniformly on any compact subset of $\mathcal{D}$.*

Proof. Using Cauchy's theorem, we have

$$f(z) = \frac{1}{2\pi i} \int_C \frac{f(t)}{t-z} \, dt, \qquad (6.1)$$

where $C = C_0 - C_1 - C_2 - \cdots - C_m$, meaning that the contour C_0 is taken counterclockwise and the other contours are taken clockwise, as shown in Figure 1.

We prove convergence on compact subsets of $\mathcal{D}$; let these be at least a distance δ from the boundary C. Clearly, $|t-z| \geqslant \delta > 0$ in (6.1).

To extract a rational approximant from the representation (6.1), we choose $\eta > 0$, we divide C into p subarcs $\Gamma_1, \Gamma_2, \ldots, \Gamma_p$, each of length at most η, and we choose arbitrary points $t_k \in \Gamma_k$. Define

$$\gamma_k = \frac{1}{2\pi i} \int_{\Gamma_k} f(t) \, dt$$

and

$$R(z, \eta) = \sum_{k=1}^{p} \frac{\gamma_k}{t_k - z}. \qquad (6.2)$$

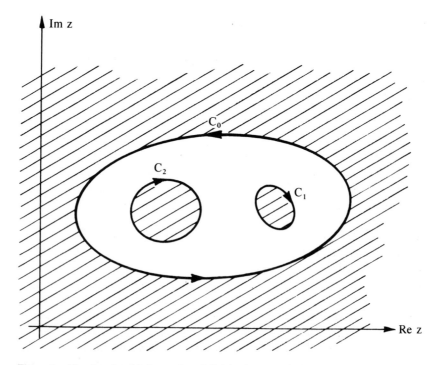

Figure 1. The domain of holomorphy of $f(z)$ is shown unshaded; the boundary curves C_0, C_1, and C_2 are also shown.

Equation (6.2) defines the rational approximant. Let $l(C)$ denote the length of C, and $M = \max|f(z)|$ over the compact subset of $\mathcal{D}$ in question. Then we have an easy error bound

$$|f(z) - R(z, m)| \leqslant \frac{Ml(C)}{2\pi\delta^2} \eta. \tag{6.3}$$

We may choose η to be arbitrarily small in (6.3); in particular, we choose $\eta \ll \delta^2 [Ml(C)]^{-1}$, and the theorem is proved.

Notice that (6.2) provides an explicit construction of the rational approximants. The next lemma generalizes Lemma 1 by allowing an infinite number of inner boundary curves which enclose nonanalytic points of $f(z)$.

LEMMA 2. *Let $\mathcal{D}$ be an arbitrary connected, open bounded domain, and let $f(z)$ be holomorphic in $\mathcal{D}$. Then a sequence of rational functions exists which converges to $f(z)$ uniformly on any compact subset of $\mathcal{D}$.*

Proof. We can exhaust $\mathcal{D}$ by a nested sequence of open domains $\{\mathcal{D}_n\}$, such that for all n

$$\overline{\mathcal{D}}_{n-1} \subset \mathcal{D}_n \subset \mathcal{D}. \tag{6.4}$$

A simple way to do this is to set up a Cartesian grid of mesh size 2^{-m} in the z-plane, as shown in Figure 2. Then let $\mathcal{E}_m$ consist of the open squares of the grid, together with their common sides, for which $\overline{\mathcal{E}}_m \subset \mathcal{D}$. We take m larger than some minimum m_0, so that $\mathcal{E}_m$ is open and connected for $m > m_0$. Then a subsequence of $\{\mathcal{E}_m, m > m_0\}$, called $\{\mathcal{D}_n\}$, exists satisfying (6.4); for each $z \in \mathcal{D}$, n_z may be found such that $z \in \mathcal{D}_{n_z}$. Further, the boundary of $\mathcal{D}_n$, called C_n, consists of simple closed polygons. Let the minimum distance δ_n of $\mathcal{D}_{n-1}$ from C_n be δ_n; we have arranged by (6.4) that $\delta_n > 0$. Using the previous lemma, we obtain rational functions $S_n(z)$ with poles on C_n such that for any $\eta_n > 0$,

$$|f(z) - S_n(z)| \leqslant \frac{M_n \lambda_n}{2\pi\delta_n^2} \eta_n,$$

where λ_n is the length of C_n and $M_n = \max|f(z)|$ for $z \in \mathcal{D}_n$. Hence λ_n, M_n are bounded. Whatever value δ_n may have, η_n may be chosen sufficiently small so that the sequence $S_n(z) \to f(z)$ uniformly on compact subsets of $\mathcal{D}$, and we may arrange that

$$|f(z) - S_n(z)| < 2^{-n}. \tag{6.5}$$

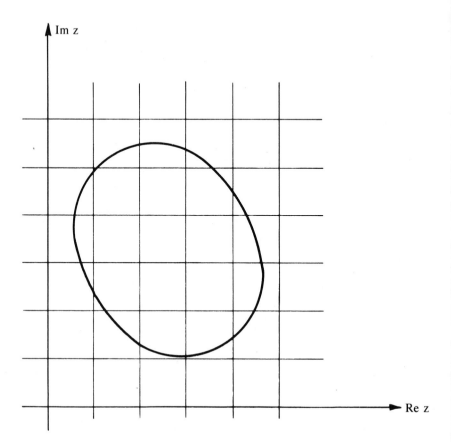

Figure 2. A Cartesian grid and the domain $\mathcal{D}$.

The construction of $S_n(z)$ in Lemma 1 consists of assigning poles on the boundary of $\mathcal{D}$, the domain of holomorphy of $f(z)$. These poles are not normally singularities of $f(z)$. The construction of Lemma 2 consists of assigning poles to $S_n(z)$ within the domain of holomorphy but outside the compact region of convergence. Runge's theorem removes this unappealing feature of the approximants by placing the poles of $S_n(z)$ at prescribed points in the excluded regions. For details, see Roth [1938] and Gauthier [1977].

THEOREM 3.6.1 [Runge, 1885]. *Let $f(z)$ be holomorphic in $\mathcal{D}$, where $\mathcal{D}$ is a bounded open domain with a complement consisting of precisely $m+1$ components, $\mathcal{F}_0, \mathcal{F}_1, \ldots, \mathcal{F}_m$. Let $\mathcal{F}_0$ contain the point $z = \infty$; then the other components $F_1, F_2, \ldots, F_m$ are compact. Let $z_k \in F_k$, $k = 0, 1, 2, \ldots, m$, be a chosen set of finite points. Then a sequence $\{R_n(z)\}$ of rational approximants exists for which $R_n(z) \to f(z)$ uniformly on compact subsets of $\mathcal{D}$, and all the poles of*

$R_n(z)$ are located at the preassigned points $\{z_k\}$. If $m=0$, the same approximation is possible taking $\{R_n(z)\}$ to be polynomials instead of rational functions.

Proof. We shall prove the case with $m \geqslant 1$. Using the method of Lemma 2, suppose that n is large enough so that the boundary of the nth polygonal curve may be expressed as

$$C_n = C_{0n} - C_{1n} - C_{2n} - \cdots - C_{mn},$$

where the exterior of C_{0n} contains $\mathscr{F}_0$ and the interiors of $C_{1n}, C_{2n}, \ldots, C_{mn}$ contain $\mathscr{F}_1, \mathscr{F}_2, \ldots, \mathscr{F}_m$ respectively. $S_n(z)$ is a rational approximant constructed to have many simple poles on each of the $m+1$ curves C_{kn}. These poles are to be replaced by precisely $m+1$ multipoles at $z_0, z_1, \ldots, z_m$ without appreciably altering the accuracy of the approximation.

To "move" a pole from $z=a$ on C_k to $z=b$ which is closer to z_k, we use the identity

$$\frac{1}{z-a} = \frac{1}{z-b} + \frac{a-b}{(z-b)^2} + \cdots + \frac{(a-b)^{p-1}}{(z-b)^p} + \frac{(a-b)^p}{(z-a)(z-b)^p},$$

so that

$$\left| \frac{1}{z-a} - \sum_{j=1}^{p} \frac{(a-b)^{j-1}}{(z-b)^j} \right| = \frac{|a-b|^p}{|z-a||z-b|^p}.$$

For any $z \in \mathscr{D}_{n-1}$ and b such that

$$2|b-a| < |z-b|, \tag{6.6}$$

we have

$$\left| \frac{1}{z-a} - \sum_{j=1}^{p} \frac{(a-b)^{j-1}}{(z-b)^j} \right| < \frac{2^{-p}}{\delta_n}. \tag{6.7}$$

The expression (6.7) is the key to the theorem, because it shows that a pole at $z=a$ may be "replaced" by a multipole of high order at $z=b$. Furthermore, any multipole located at $z=a$ [i.e. a polynomial in $(z-a)^{-1}$] may be similarly approximated. If we could take $b=z_k$, the theorem would now be proved, but in general (6.6) does not hold with $b=z_k$. Instead, we must find a sequence of points on a polygonal line L joining a to z_k, given by

$$b_0 = a, \; b_1, \; b_2, \ldots, \; b_N = z_k,$$

such that $|b_j - b_{j-1}| < \frac{1}{2}\delta$, $j = 1, 2, \ldots, N$, where δ is the distance from L to $\mathfrak{D}_{n-1}$. This condition maintains the inequality

$$|b_j - b_{j-1}| < \frac{1}{2}|z - b_j|.$$

Now, by expansion and reexpansion, but with a finite total number of steps, we may find $R_n(z)$ with poles at $z_0, z_1, \ldots, z_k$ such that

$$|S_n(z) - R_n(z)| < 2^{1-n}. \tag{6.8}$$

Equations (6.5), (6.8) together show that $|f(z) - R_n(z)| < 2^{1-n}$ for $z \in \mathfrak{D}_{n-1}$, completing the proof for $m > 0$. We leave the proof for $m = 0$ as a straightforward exercise, [Hille, 1962, p 299].

It is interesting to contrast Runge's theorem with Laurent's theorem about convergent rational approximation within an annulus of holomorphy. In this case, there are two multipoles, one in the annulus and one at infinity.

Thus far, we have established the possibility of obtaining uniform convergence with rational approximants to a function with a domain of holomorphy such as Figure 1 or any Hepworth sculpture. We are a long way from being able to prove that even a subsequence of Padé approximants converges in such a domain. However, we can easily show that polynomial approximation is certain cases is futile, and why Runge's theorem is so important.

We will show that $p_n(z) = 0$ is a best polynomial approximation of order n to $f(z) = z^{-1}$ on the annulus $1 \leqslant |z| \leqslant 2$. For suppose that $\tilde{p}_n(z)$ is a better polynomial approximation; then

$$\max_{1 \leqslant |z| \leqslant 2} |\tilde{p}_n(z) - z^{-1}| = r < 1. \tag{6.9}$$

By Cauchy's theorem

$$\int_{|z|=1} \left\{ \tilde{p}_n(z) - z^{-1} \right\} dz = -2\pi i.$$

Hence

$$\int_{|z|=1} |\tilde{p}_n(z) - z^{-1}| \, |dz| \geqslant 2\pi,$$

which contradicts (6.9). Thus there is no better polynomial approximation (in the variable z) than $p_n = 0$ to z^{-1} on the annulus; of course, a rather simple rational approximant is exact.

The following two theorems show the role of Padé approximants in the theory of best rational approximation in the Tchebycheff sense. The first theorem we state without proof. It shows the similarity between the convergence properties of Padé approximants in the ideal conditions under which de Montessus's theorem holds, and best rational approximations under the same conditions.

THEOREM 3.6.2 [Walsh, 1964a]. *Let $f(z)$ be analytic in the disk $|z| < r$ and be meromorphic with precisely M poles, counting multiplicity, in the larger disk $|z| < \rho$. If $0 < \varepsilon < r$, then the rational functions $R^{(L/M)}(\varepsilon, z)$ of type (L/M) of best approximation to $f(z)$ on the closed disk $|z| \leq \varepsilon$ converge as $L \to \infty$ uniformly to $f(z)$ on any compact subset of $|z| < \rho$ containing no poles of $f(z)$. Also, the poles of $R^{(L/M)}(\varepsilon, z)$ converge to the M poles of $f(z)$ lying in $r < |z| < \rho$.*

We omit the proof, because the theorem is only indirectly connected with Padé approximants. Next, we both state and prove another theorem which shows that Padé approximants are the limits of best rational approximants in a vanishingly small neighborhood of the origin.

THEOREM 3.6.3 [Walsh, 1964b]. *Let $f(z)$ be analytic at the origin, and let $R^{(L/M)}(\varepsilon, z)$ be the best rational approximant of type (L/M) in $|z| \leq \varepsilon$. Provided $C(L/M) \neq 0$, then as $\varepsilon \to 0$, $R^{(L/M)}(\varepsilon, z) \to [L/M]_f(z)$ uniformly on any compact set containing no pole of $[L/M]_f(z)$.*

Proof. Because $R^{(L/M)}(\varepsilon, z)$ is the best rational approximant of $f(z)$, it has no pole in $|z| \leq \varepsilon$ for some preassigned positive ε, and therefore

$$R^{(L/M)}(\varepsilon, z) = \sum_{k=0}^{\infty} r_k(\varepsilon) z^k \qquad \text{for} \quad |z| < \varepsilon.$$

Using the contour representation, we further know that

$$r_k(\varepsilon) - c_k = \frac{1}{2\pi i} \int_{|z|=\varepsilon} \frac{R^{(L/M)}(\varepsilon, z) - f(z)}{z^{k+1}} dz.$$

Again, because $R^{(L/M)}(\varepsilon, z)$ is the best rational approximant on $|z| \leq \varepsilon$,

$$\left| R^{(L/M)}(\varepsilon, z) - f(z) \right| < \left| [L/M]_f(z) - f(z) \right| \qquad \text{for} \quad |z| = \varepsilon,$$

and so

$$|r_k(\varepsilon) - c_k| \leq \frac{1}{2\pi} \int_{|z|=\varepsilon} \frac{|[L/M]_f(z) - f(z)|}{|z|^{k+1}} |dz|$$

$$= O(\varepsilon^{L+M-k+1}).$$

Consequently $r_k(\varepsilon) \to c_k$ as $\varepsilon \to 0$ for $k = 0, 1, \ldots, L + M$. The coefficients $\{c_k, k = 0, 1, \ldots, L + M\}$ uniquely determine $[L/M]$ if $C(L/M) \neq 0$. Provided $|r_k(\varepsilon) - c_k|$ is sufficiently small for $k = 0, 1, \ldots, L + M$, $R^{(L/M)}(\varepsilon, z)$ is uniquely determined by $r_k(\varepsilon)$ for these values of k, and $R^{(L/M)}(\varepsilon, z) \to [L/M]_f(z)$ uniformly if the respective denominators do not vanish; these requirements are met by the conditions of the theorem.

This theorem of Walsh is important because it may be interpreted as the statement that Padé approximants are the best local rational approximants. There are two interesting related results, which we present as corollaries without proof.

COROLLARY 1. *Theorem 3.6.3 holds good when $R^{(L/M)}(\varepsilon, z)$ is defined as the best rational approximation on the real closed interval $0 \leq z \leq \varepsilon$ (instead of the disk $|z| \leq \varepsilon$).*

COROLLARY 2. *Let $f(z)$ be a real-valued function with $m + n + 1$ continuous derivatives on $[0, \delta]$ for some $\delta > 0$. For each ε, $0 < \varepsilon < \delta$, let $R^{(L/M)}(\varepsilon, z)$ be the best (L/M)-type approximant to $f(z)$ on $[0, \varepsilon]$. Then $R^{(L/M)}(\varepsilon, z) \to [L/M]_f(z)$ as $\varepsilon \to 0$ on some closed interval $[0, \varepsilon_0]$, $0 < \varepsilon_0 \leq \delta$, using the classical definition, not the Baker definition, of $[L/M]_f(z)$.*

Corollary 1 is simply a result different from but similar to Walsh's theorem. Corollary 2 shows that Padé approximants with the classical definition emerge as the limit of best local rational approximation [Chui et al., 1974].

For further details of recent progress with Walsh-type theorems, we refer to Chui et al. [1975], Chui et al. [1978], and Walsh [1965a, b, 1970]. Examples which show the nonuniqueness of best rational complex-valued approximations to real functions on real intervals [Saff and Varga, 1978] are cautionary and interesting. A good general reference on rational approximation per se is Hille [1962].

3.7 Padé Approximants for the Riccati Equation

A Riccati equation is a nonlinear ordinary differential equation of the form

$$A(x) \frac{du}{dx} + B(x)u + C(x)u^2 = D(x). \tag{7.1}$$

Riccati equations arise naturally in the analysis of homogeneous, second-order, ordinary differential equations. For example, we consider the one-dimensional Schrödinger equation, which may be expressed as

$$\phi'' - D(x)\phi = 0. \tag{7.2}$$

By making the substitution $u(x)=\phi'(x)/\phi(x)$, (7.2) becomes

$$u'+u^2=D(x),$$

which is a special case of (7.1). In fact, the general Riccati equation of the form (7.1) can always be reduced to a second-order, linear homogeneous ordinary differential equation [Ince, 1944, p. 24]. If the functions $A(x)$, $B(x)$, $C(x)$, and $D(x)$ occurring in (7.1) are polynomials, it may be possible to solve (7.1) using a continued-fraction variant of the Frobenius method. Notice that the substitution

$$u(x)=\frac{1}{f(x)}$$

in (7.1) leads to the equation

$$A(x)f'-B(x)f+D(x)f^2=C(x), \tag{7.3}$$

which is also a Riccati equation. In this sense, the Riccati equation is "form invariant under reciprocation".* We explain a method of solution by means of an illustrative example. Consider the Riccati equation

$$(ax+b)y'+(cx+d)y+(ex+g)y^2=h, \tag{7.4}$$

where a, b, c, d, e, g, and h are constants. Substitute

$$y(x)=\frac{h}{cx+\alpha-u(x)}, \tag{7.5}$$

which leads to the equation

$$(ax+b)u'+(cx+\alpha-u)(d-\alpha+u)+h(ex+g)-c(ax-b)=0. \tag{7.6}$$

Equations (7.5), (7.6) are consistent with the hypothesis that

$$u(x)=O\!\left(\frac{1}{x}\right), \tag{7.7}$$

as $|x|\to\infty$, provided that the value of α is taken to be

$$\alpha=d-a+eh/c. \tag{7.8}$$

Then (7.6) reduces to

$$(ax+b)u'+(cx-d+2\alpha)u-u^2=\alpha(\alpha-d)-gh+cb. \tag{7.9}$$

*J. Zinn-Justin, private communication.

Comparison of (7.4) and (7.9) shows that this Riccati equation is form invariant under the substitution (7.5). The constants a, b, and c are unchanged, and

$$d \to 2\alpha - d, \quad e \to 0, \quad g \to -1, \quad h \to \alpha(\alpha - d) - gh + cb. \quad (7.10)$$

A solution of an arbitrary Riccati equation of the type (7.4) can therefore be expressed as a continued fraction by making repeated substitutions of the form (7.5). As an example of a Riccati equation of this kind, we consider

$$xy' - (x + n - 1)y = -1. \quad (7.11)$$

Continued-fraction solutions of differential equations of this type were first found by Laguerre [1879, 1885], who used the equation (7.11) in the case $n = 1$ as an explicit example. Equation (7.11) has the solution

$$y(x) = e^x E_n(x), \quad (7.12)$$

where

$$E_n(x) = \int_1^\infty e^{-xt} t^{-n} \, dt = x^{n-1} \int_x^\infty e^{-u} u^{-n} \, du \quad (7.13)$$

for $\operatorname{Re} x > 0$. Following (7.5), (7.8), we substitute

$$y(x) = \frac{-1}{-x - n - u(x)}, \quad (7.14)$$

and from (7.10) we find that $u(x)$ satisfies

$$xu' - (x + n + 1)u - u^2 = n. \quad (7.15)$$

Now we consider the inductive hypothesis

$$xu_r' - (x + n + 2r - 1)u_r - u_r^2 = r(n + r - 1), \quad (7.16)$$

and note that (7.15) is precisely (7.16) in the case $r = 1$. Using (7.5), we substitute

$$u_r(x) = \frac{r(n + r - 1)}{-x + \alpha_r - u_{r+1}(x)}. \quad (7.17)$$

Using (7.8), we find that

$$\alpha_r = -n - 2r \quad (7.18)$$

and from (7.10) we find that $u_{r+1}(x)$ satisfies

$$xu'_{r+1} - (x+n+2r+1)u_{r+1} - u^2_{r+1} = (r+1)(n+r). \qquad (7.19)$$

Equation (7.19) establishes the inductive hypothesis (7.16) for all positive integer r. We find the solution of (7.11) from (7.14), (7.17), and (7.18) to be

$$y(x) = \frac{-1}{-x-n} - \frac{n}{-x-n-2} - \frac{2n+2}{-x-n-4} - \cdots$$
$$\cdots - \frac{(r+1)(n+r)}{-x-n-2(r+1)} - \cdots$$
$$= \frac{1}{x+n} - \frac{n}{x+n+2} - \frac{2n+2}{x+n+4} - \cdots - \frac{(r+1)(n+r)}{x+n+2(r+1)} - \cdots.$$

$$(7.20)$$

In terms of the variable $z = x^{-1}$, the Mth convergent of (7.20) is the $[M/M]$ Padé approximant of $y(1/z)$, by virtue of the accuracy-through-order conditions imposed by (7.5), (7.7), and (7.8). In fact the fraction (7.20) is precisely that of (I.4.6.8), although the method of derivation is quite different. The foregoing analysis makes it clear that neat, closed-form expressions for the solution of Riccati equations can only be obtained in exceptional cases. In general, the technique exemplified by (7.4), (7.5), (7.8), and (7.10) leads to a continued-fraction representation of $y(x)$, the solution of the Riccati equation, provided the accuracy-through-order conditions can be imposed consistently. A necessary condition for this consistency in the case that the coefficients $A(x)$, $B(x)$, $C(x)$, and $D(x)$ occurring in (7.1) are polynomials is that the degrees of these polynomials satisfy

$$\deg\{D(x)\} < \deg\{B(x)\},$$
$$\deg\{A(x)\} \leqslant \deg\{B(x)\},$$
$$\deg\{C(x)\} \leqslant \deg\{B(x)\}.$$

For further details of the theory of Riccati equations, we refer to Reid [1972]; for their connection with the Lanczos τ-method, we refer to Fair [1964]; and for an application to Laguerre's method to the expansion of the Hamburger function

$$G_\beta(x) = \int_{-\infty}^{\infty} \exp\left(\frac{-t^2/2 - \beta t^4}{x-t}\right),$$

we refer to Bessis [1979].

CHAPTER 4

Connection with Quantum Field Theory

4.1 Perturbed Harmonic Oscillators

This chapter consists of three sections. In the first section, we discuss the arguments which lead us to think that Padé approximants are an important tool for the summation of the perturbation expansions of quantum field theories. In the second section, we discuss two exemplary models of pion–pion scattering in which Padé approximants have revitalized the prospects of perturbation theory. In the last section we give a flavor of the application of Padé methods to critical phenomena and indicate how this subject relates to field theory.

We cannot adequately summarize here the many books on quantum theory. The relevant background of quantum mechanics used in this section is given in "The Principles of Quantum Mechanics" [Dirac, 1958]. Elementary field theory and quantum electrodynamics are described by Schweber [1961]; we refer especially to the "Cargèse Lectures in Physics" [Bessis, 1972] for an advanced treatment of the relevant background of quantum field theory.

The world of quantum electrodynamics (QED) is the world of photons and electrons. The vacuum is postulated to be the lowest energy state. A basic feature of QED is the possibility of the formation of virtual electron–positron pairs. This phenomenon is the origin of vacuum polarization, which has a measurable and accurately verified effect on the energy levels of atomic hydrogen. Exploding black holes may be a more dramatic realization of this effect. Dyson [1952] considered a configuration of many nearby electrons and, some distance away, as many nearby positrons. We show an

ENCYCLOPEDIA OF MATHEMATICS and Its Applications, Gian-Carlo Rota (ed.). Vol. 14: George A. Baker, Jr., and Peter R. Graves-Morris, Padé Approximants, Part II: Extensions and Applications   ISBN 0-201-13513-2

Figure 1. A configuration of virtual electrons and positrons.

example in Figure 1. The self-interaction energy of such a virtual system is dominated by the short-range repulsion energy of each group. Thus the configuration has a large positive self-energy, corresponding to a very small temporal duration and little effect on any calculation.

Suppose that any quantum-electrodynamic quantity is calculated from a perturbation series in $\lambda = e^2$, and that this series converges. Then its radius of convergence is at least e^2, and so it also converges for $\lambda = -\frac{1}{2} e^2$. This corresponds physically to reversing the sign of the electron–positron inter-action. The configuration of Figure 1 now has a large negative energy and becomes exceedingly probable. This physical effect is imaginatively called the exploding vacuum. It should be represented by a divergence of any perturbation calculation in QED. If any series expansion of QED is to be associated with a function analytic in $|\lambda| < \varepsilon$, $|\arg(\lambda)| < \pi$, we can only conclude that such a series expansion should have zero radius of conver-gence.

We can illustrate this physical argument by two models which support the contention that Padé approximants or similar techniques will have to be used to interpret the perturbation series. These are quantum-mechanical harmonic-oscillator models which are analogues of a zero-dimensional field theory.

The Peres Model

Peres considered two independent one-dimensional quantum-mechanical harmonic oscillators. The mth excited state of one is described by the wave function

$$|m\rangle_1 = \psi_m(x) = c_m e^{-\frac{1}{2}x^2} H_m(x), \tag{1.1a}$$

and the nth excited state of the other by

$$|n\rangle_2 = \psi_n(y) = c_n e^{-\frac{1}{2}y^2} H_n(y). \tag{1.1b}$$

where c_m, $m = 0, 1, 2, \ldots$, are normalization constants. The combined wave function of the oscillators is

$$|m, n\rangle = \psi_{m,n}(x, y) = c_m c_n e^{-\frac{1}{2}(x^2 + y^2)} H_m(x) H_n(y), \tag{1.2}$$

which is an eigenfunction of the Hamiltonian

$$\mathcal{K}_0 = \tfrac{1}{2}\left(p_x^2 + p_y^2 + x^2 + y^2 \right).$$ (1.3)

The time independent solution (1.2) is perturbed by the instantaneous interaction

$$\mathcal{K}_I = gx^2 y\delta(t).$$ (1.4)

Its effect is determined by the time dependent Schrödinger equation

$$\{\mathcal{K}_0 + \mathcal{K}_I\}\Psi(x, y, t) = i\frac{\partial\Psi(x, y, t)}{\partial t}$$

For $t < 0$, the solutions are the unperturbed solutions

$$\Psi_{m,n}(x, y, t) = \psi_{m,n}(x, y)e^{-i(E_m + E_n)t}$$

and so for $t = 0+$, the perturbed wave function is

$$\Psi_{m,n}(x, y, 0+) = e^{-igx^2 y}\Psi_{m,n}(x, y, 0-).$$

The S-matrix for the interaction is therefore given by the overlap integral

$$S(m'n', mn) = \int_{-\infty}^{\infty}\int_{-\infty}^{\infty} \psi^*_{m',n'}(x, y)e^{-igx^2 y}\psi_{m,n}(x, y)\,dx\,dy.$$ (1.5)

The ground-state wave function, given by (1.1) and normalized, is

$$\psi_{0,0}(x, y) = \frac{1}{\sqrt{\pi}}e^{-\tfrac{1}{2}(x^2 + y^2)}.$$

Substituting in (1.5), the ground-state-to-ground-state transition amplitude is

$$S_{00}(g^2) = \frac{1}{\pi}\int_{-\infty}^{\infty}\int_{-\infty}^{\infty} e^{-x^2 - y^2 - igx^2 y}\,dx\,dy$$ (1.6)

$$= \frac{2}{\sqrt{\pi}}\int_0^{\infty} e^{-x^2 - \tfrac{1}{4}g^2 x^4}\,dx$$ (1.7)

$$= \pi^{-1/2}g^{-1}e^{1/(2g^2)}K_{1/4}\left(\frac{1}{2g^2}\right).$$ (1.8)

$S_{00}(g^2)$ is well defined by (1.6), (1.7), or (1.8) for all real values of g. Further, (1.7) shows that $S_{00}(g^2) \to 1$ as $g \to 0$ for real g. However, (1.7) is undefined for $\mathrm{Re}\,g^2 < 0$, and it is plain that $S_{00}(g^2)$ has a singularity at

$g^2 = 0$. While (1.8) is an explicit expression for $S_{00}(g^2)$, it is more convenient to integrate over x first in (1.6) and obtain

$$S_{00}(g^2) = \frac{1}{\sqrt{\pi}} \int_{-\infty}^{\infty} e^{-y^2} \frac{1}{\sqrt{1+igy}} \, dy$$

$$= \frac{1}{\sqrt{\pi}} \int_{-\infty}^{\infty} e^{-y^2} \left\{ 1 + \frac{(-\frac{1}{2})(-\frac{3}{2})}{2!} (-g^2 y^2) \right.$$

$$\left. + \frac{(-\frac{1}{2})(-\frac{3}{2})(-\frac{5}{2})(-\frac{7}{2})}{4!} (-g^2 y^2)^2 + \cdots \right\} dy$$

$$= \frac{1}{\sqrt{\pi}} \int_{-\infty}^{\infty} e^{-y^2} \left\{ 1 + \frac{(\frac{1}{4})(\frac{3}{4})}{(\frac{1}{2})1!} (-g^2 y^2) \right.$$

$$\left. + \frac{(\frac{1}{4})(\frac{3}{4})(\frac{5}{4})(\frac{7}{4})}{(\frac{1}{2})(\frac{3}{2})2!} (-g^2 y^2)^2 + \cdots \right\} dy.$$

Using the recurrence relation for the gamma function and

$$\int_0^{\infty} e^{-y^2} y^{2r} \, dy = \Gamma(r + \tfrac{1}{2}),$$

we obtain

$$S_{00} = 1 + \frac{(\frac{1}{4})(\frac{3}{4})(-g^2)}{1!} + \frac{(\frac{1}{4})(\frac{3}{4})(\frac{5}{4})(\frac{7}{4})}{2!} (-g^2)^2 + \cdots \qquad (1.9)$$

$$= {}_2F_0\left(\tfrac{1}{4}, \tfrac{3}{4}, -g^2\right). \qquad (1.10)$$

These formal manipulations with divergent series are justified by the use of
Carleman's theorem. Also we see that $S_{00}(g^2)$ is a Stieltjes series with zero
radius of convergence. The asymptotic series (1.9) shows that finite-order
perturbation theory with sufficiently weak coupling can give reasonably
accurate results. However, arbitrarily accurate results for any specified
coupling cannot be obtained by simple series summation. Diagonal Padé
approximants to (1.9) are known to converge, and numerical calculations
confirm this impressively [Baker and Chisholm, 1966].

We review this analysis using the Heisenberg representation, in which the
space variable x and the momentum p_x are time dependent operators. They
satisfy the equal-time commutation relation

$$[x(t), p_x(t)] = i.$$

The solution is given by

$$x(t) = x_0 \cos t + p_{x0} \sin t,$$
$$p_x(t) = p_{x0} \cos t - x_0 \sin t,$$

and it follows that

$$[x(t), x(t')] = -i \sin(t - t').$$ (1.11)

We consider next the operators

$$\eta = \frac{p + ix}{\sqrt{2}}$$

and

$$\eta^* = \frac{p - ix}{\sqrt{2}},$$

which play the roles of creation and annihilation operators. Following Dirac, we find that the Heisenberg energy eigenstates are given by $\eta^n |0\rangle$ with eigenvalues $E_n = n + \frac{1}{2}$. Then $x = (\eta - \eta^*)/\sqrt{2}$ corresponds to the quantum field, and its propagator is given by (1.11). Hence we see that the interaction Hamiltonian (1.4) is analogous to a $g\bar{\psi}\psi\phi$ interaction, and similarly to the electron–photon interaction of QED. We are led to expect a singularity at zero coupling strength in the QED perturbation expansions.

The Anharmonic Oscillator

The quantum-mechanical one-dimensional anharmonic oscillator considered here is derived from the Hamiltonian

$$\mathcal{H} = p^2 + x^2 + \beta x^4, \qquad \beta > 0.$$ (1.12)

The quantum-mechanical problem consists of finding the energy eigenvalues and eigenstates of the Schrödinger equation

$$-\frac{d^2\psi}{dx^2} + (x^2 + \beta x^4)\psi = E\psi.$$ (1.13)

If $\beta = 0$, the solution is given by (1.1a) and perturbation theory provides a method for calculating E and $\psi(x)$ as power series in β.

The leading term of the asymptotic solution of (1.13) is

$$\psi(x) \sim e^{-\sqrt{\beta} x^3/3} \qquad \text{as} \quad x \to +\infty.$$

We see that the character of the wave function changes drastically at $\beta=0$. It is better to consider the alternative problem with

$$H=p^2+\alpha x^2+x^4$$

and solve the Schrödinger equation

$$-\frac{d^2\psi}{dx^2}+(\alpha x^2+x^4)\psi=E\psi.$$

For this, the perturbative solutions can be proved to be analytic in α at $\alpha=0$, and the expansion converges for small $|\alpha|$. More generally, we consider the Hamiltonian

$$H=p^2+\alpha x^2+\beta x^4, \tag{1.14}$$

which has energy levels $E=E_n(\alpha,\beta)$. We have stated that $E_n(\alpha,1)$ is analytic at $\alpha=0$.

Consider the scale change in (1.13) of

$$x\rightarrow x'=\lambda x.$$

Any one solution of (1.13) leads to a whole class of solutions of (1.14), and we find

$$E_n(\alpha,\beta)=\lambda^{-2}E_n(\alpha\lambda^4,\beta\lambda^6).$$

By taking $\alpha=1$, $\beta\lambda^6=1$, this becomes

$$E_n(1,\beta)=\beta^{1/3}E_n(\beta^{-2/3},1), \tag{1.15}$$

which shows that $E_n(1,\beta)$ has a three-sheeted branch cut ending at ∞ in the β-plane. Further, it suggests strongly (and correctly) that the three-sheeted branch point is at $\beta=0$. Since $E_n(1,\beta)$ is real symmetric, the cut is taken along $-\infty<\beta\leqslant 0$, and we emphasize the consequence that the perturbation series of $E_n(1,\beta)$ has zero radius of convergence. Again, it can be proved that $E_n(1,\beta)$ is a Stieltjes series in β with a convergent sequence of diagonal Padé approximants. For further details of the theory of the anharmonic oscillator and its relation to field theory, we refer to Bender and Wu [1968] and Simon [1970]. The paper of Loeffel et al. [1969] is important for proving the convergence of the Padé method for the βx^4 anharmonic oscillator. For the βx^8 anharmonic oscillator, the perturbation series diverge so rapidly that the moment problem is indeterminate [Graffi and Grecchi, 1978]. For further details, we refer to Killingbeck [1980], Graffi et al. [1971], Graffi et al. [1970], and Ruijgrok [1972].

In the Heisenberg picture, the βx^4 perturbation in (1.12) corresponds to ϕ^4 field theory. The inference is that the perturbation expansion of field theory is always divergent and that Padé approximants should be used. Indeed, recent explicit estimates show that the perturbation series coefficients in four dimensions diverge like

$$c_k \sim \left(\frac{-1}{16\pi^2}\right)^k \left(k + \tfrac{7}{2}\right)! \qquad \text{as} \quad k \to \infty$$

[Brezin et al., 1977].

4.2 Pion-Pion Scattering

In this section we discuss some applications of Padé approximants by which fundamental theories of elementary particles are tested. We concentrate on how the properties of the Padé approximants are useful in these applications rather than on the details of the dynamics. No longer are we in a situation of having arbitrarily many terms of a power series available to arbitrary accuracy. With pion dynamics, at most five terms are available at present, and laborious calculation is required to obtain even a few decimal places of accuracy of the highest-order term. Space does not permit a full discussion of the technicalities of elementary-particle theory in this section. Again we refer to the "Cargèse Lectures in Physics" [Bessis, 1972] for details.

A variety of calculations have been made using Padé approximants on nucleon–nucleon and meson–nucleon systems. The prototype Padé approximant calculation of the hadronic spectrum is that of the ϕ^4 model of pion–pion scattering. The perturbation series is derived from the Lagrangian

$$L = \tfrac{1}{2}\left(\partial_\mu \phi_\alpha\right)^2 + \tfrac{1}{2}m^2\phi_\alpha^2 + \tfrac{1}{4}\lambda\left(\phi_\alpha\phi_\alpha\right)^2. \tag{2.1}$$

The T-matrix elements are derived from the Feynman graphs of Figure 1, and formally

$$T(s,t,u) = \sum_{i=0}^{\infty} T_i(s,t,u).$$

Each independent closed loop of each graph represents a four-dimensional space-time integral, and each internal line represents a propagator. A factor of λ is associated with each four-line vertex. If the graph represents a scattering process in which the intermediate particles can satisfy classical energy and momentum conservation, as if the pions were colliding billiard balls, then the graphs are singular and necessarily require multidimensional

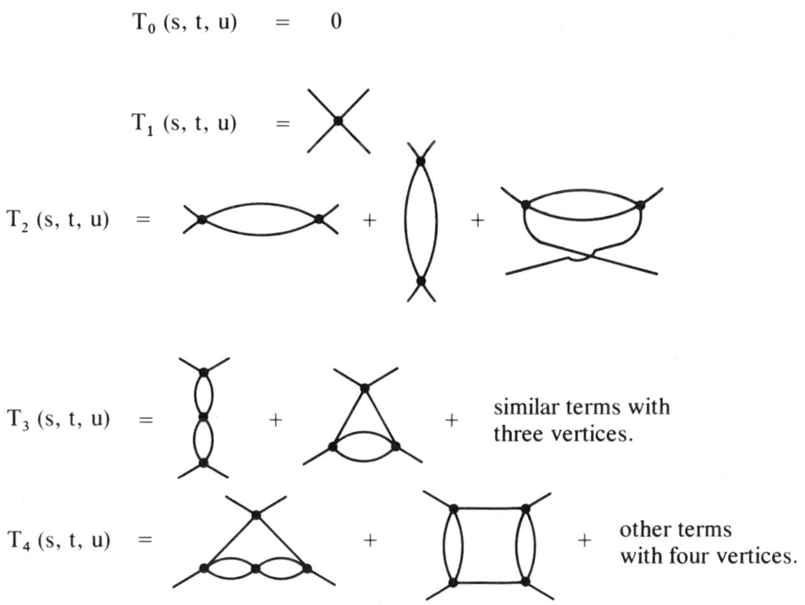

$$T_0\,(s,\,t,\,u) \quad = \quad 0$$

$$T_1\,(s,\,t,\,u) \quad = \quad \text{}$$

$$T_2\,(s,\,t,\,u) \quad = \quad \text{} \quad + \quad \text{} \quad + \quad \text{}$$

$$T_3\,(s,\,t,\,u) \quad = \quad \text{} \quad + \quad \text{} \quad + \quad \text{similar terms with three vertices.}$$

$$T_4\,(s,\,t,\,u) \quad = \quad \text{} \quad + \quad \text{} \quad + \quad \text{other terms with four vertices.}$$

Figure 1. Some Feynman graphs.

singular quadratures. Furthermore, such graphs are always numerically significant, so that accurate evaluation of them is important.

From an analysis of the theory of Feynman graphs, it is known that the T-matrix is crossing symmetric, which means that $T(s, t, u)$ is a symmetric function of all three variables s, t, and u. The variable s corresponds to the energy available for the scattering process, and t to the momentum transfer. s, t, and u are relativistic invariants formed from the pions' four-momenta, and $s+t+u=4$; consequently, u is a dependent variable. Ignoring the complication of isospin, the perturbation series of the T-matrix is crossing symmetric order by order:

$$T_i(s, t, u) = T_i(t, u, s) = T_i(u, t, s). \tag{2.2}$$

Consequently formation of any $[L/M]$ Padé approximant to the T-matrix preserves crossing symmetry.

Another important characteristic of the T-matrix is its unitary property. We form partial waves by making the Legendre decomposition

$$t_l(k^2) = \rho(k^2) \int_{-1}^{1} T\big(4+4k^2, \, -2k^2(1-x), \, -2k^2(1+x)\big) P_l(x)\,dx,$$

where $\rho(k^2)$ is a kinematical factor. k is the modulus of momentum of any of the pions in the center-of-mass frame, and $x=\cos\theta$ is the cosine of the scattering angle (θ). The unitary property of the T-matrix is expressed by

$$\left[1+it_l(k^2)\right]\left[1+it_l(k^2)\right]^* = 1. \tag{2.3}$$

Just as in potential theory (2.4.12), any finite order of perturbation theory is not exactly unitary. However, diagonal approximants of the S-matrix and $[L/M]$ approximants of the T-matrix perturbation series are unitary, provided $M \geq L$ (see Part I, Section 1.5: Theorem 1.5.5 and Exercise 1).

We summarize the situation by noting that all Padé approximants of the full amplitude are crossing symmetric but not unitary, whereas diagonal approximants of the partial-wave amplitudes are unitary but are not components of a crossing symmetric amplitude. We are faced with a dilemma about which is the best method to choose. In fact, it is best to use both methods and compare them for stability and consistency.

So far, it has been tacitly assumed that Padé approximants are used to accelerate convergence of the perturbation series, but there is also a strong additional reason for forming the approximants, namely their suitability for representing resonances. Many resonances occur in the various channels of pion elastic scattering, such as the ρ-meson in the $I=J=1$ channel, where I denotes the isotopic spin and J denotes the angular momentum of the pions. If the $I=J=1$ channel is open, the ρ-meson is the dominant feature of the pion–pion interaction at center-of-mass energy near 760 MeV, corresponding to $k^2 \approx 6.4$ in pion mass units.

It is widely thought that the existence of this meson is a consequence of an effective pion–pion Lagrangian, such as (2.1). Hence the ρ-meson would correspond to a singularity of the amplitude computed from the Lagrangian, and in fact the ρ-meson would be a pole of the partial-wave amplitude. More precisely, this means that $t_l(k^2)$ with g constant has a pole near $k^2 = 6.4$. However, it is natural to think of the Padé approximant as generating a pole in the coupling strength λ^2 at fixed k^2. In fact the pole of t_l in λ^2 at fixed k^2 corresponds to two second-sheet poles of $t_l(k^2)$ at fixed λ^2 if unitarity and analyticity are properly satisfied.

We are now in a position to assess qualitatively the achievements of theories based on a $\lambda\phi^4$ Lagrangian. The $I=J=1$ amplitude should contain the ρ pole, and we assume that the coupling strength is adjusted so that a particular Padé approximant of the partial-wave perturbation series has a pole at the right energy. The choice of which approximant to use is rather restricted because the constant and first-order terms of the λ^2 expansion of $t_1^1(k^2)$ are zero. Bessis and Pusterla [1967] used the [2/2] approximant, calculated λ^2 from the ρ mass, and thereby fixed the only free parameter of the theory. The predictions of the theory are the partial-wave phase shifts, the resonances, and their widths. These quantities should satisfy stability tests, such as comparison with the results from [2/1] and [1/2] approximants, and the whole model should not violate crossing symmetry too much. In short, the model predictions are that the ρ-meson is clearly present and stable, but much too narrow; the $f^0(I=0, J=2)$ is predicted but much too narrow; the S-waves are badly wrong; and an exotic ($I=2, J=2$) meson is incorrectly predicted. Crossing symmetry is tested by the Martin inequalities and is well satisfied.

The alternative approach, using Padé approximants of the full amplitude, requires isolation of the "pole term". Rather imprecisely, we expect

$$[2/2]_T(k^2, \cos\theta) = \frac{P_1(\cos\theta)}{k^2 - 6.4 - i\Gamma k} \cdot \text{constant} + \text{background}.$$

The denominator of the Padé approximant is a function of (s, t) or equivalently $(k^2, \cos\theta)$. The zero of the denominator defines the poloid, as it is called, and this should occur at a value of k^2 which is independent of $\cos\theta$. The flatness of the poloid in the physical region, $-1 \leqslant \cos\theta \leqslant 1$, is an important consistency test of this approach, which is well enough satisfied in the $\lambda\phi^4$ fourth-order model. With this approach, the partial-wave amplitudes are defined by

$$t_l(k^2) = \rho(k^2) \int_{-1}^{1} P_l(x)[2/2]_T\big(4 + 4k^2, -2k^2(1-x), -2k^2(1+x)\big)\, dx.$$

These should be unitary in the sense that (2.3) is approximately satisfied. Again, violation by the $\lambda\phi^4$ model is only a few per cent.

We may summarize the results of the $\lambda\phi^4$ theory using low-order Padé approximants of either the full amplitude or the partial-wave amplitudes by noting that the internal-consistency tests are well satisfied and that only some of the predictions of the model are correct. These results were convincing enough to stimulate considerable further work, but also indicated that the $\lambda\phi^4$ Lagrangian is probably incomplete. We do not consider here the deeper question of whether the results of the renormalized perturbation series for $\lambda\phi^4$ theory in four dimensions in fact represent the true results of the field theory. In two and three dimensions, the perturbation series is Borel summable to the Schwinger functions [Eckmann et al., 1974; Dimock, 1974; Magnen and Sénéor, 1976; Feldman and Osterwalder, 1976].

The best physical predictions of pion–pion scattering have come from the σ-model. The method is basically the same as the one described, except that the Lagrangian is

$$\mathcal{L} = \tfrac{1}{2}\big[(\partial_\mu\sigma)^2 + (\partial_\mu\pi_\alpha)^2\big] - \tfrac{1}{2}\mu^2\big[\sigma^2 + \pi_\alpha^2\big] - \tfrac{1}{4}\lambda\big[\sigma^2 + \pi_\alpha^2\big]^2 + c\sigma. \quad (2.14)$$

We have introduced a scalar field σ which is a two-pion ($I=0$, $J=0$) resonance. The Lagrangian has chiral symmetry, only broken by the $c\sigma$ term. There are two expansion parameters, c and λ; their values are fixed by the ρ mass and the pion decay rate (through partial conservation of the axial-vector current). The order of the graphs in the perturbation expansion is determined by the number of loops, enabling ordinary one-variable Padé approximants to be used. Regrettably, the model was only calculated in the one-loop approximation, permitting formation of the $[1/1]$ Padé approximant. However, even in this low order, the $I=0$ and $I=2$ low-energy S-wave phase shifts are in good agreement with experiment, which is remarkable. Furthermore, the good features of the $\lambda\phi^4$ model, namely the existence of ρ and f^0 mesons, are retained, and the consistency tests are well satisfied. On the debit side, there are insufficient terms of the series for stability tests, the $I=2$ exotic meson is still predicted, and the ρ and f^0 mesons are too narrow. Even so, the most exciting thing about the σ-model is that the best features of the $\lambda\phi^4$ theory are retained and the worst feature of the $\lambda\phi^4$ theory—the bad S-waves—have been corrected.

Just which ingredients of the complete correct theory are lacking in (2.4) is not at all clear. Undoubtedly, the status of the σ-model could be clarified using more modern techniques, such as using the off-shell momentum as a variational parameter [Alabiso et al., 1972a, b], and two-variable approximants [Graves-Morris and Samwell, 1975]. This penultimate remark reflects the theme of this work—how to make the most of numerical power series.

For further details of the status of the model described, we refer to the reviews of Basdevant [1970, 1973] and Zinn-Justin [1970], and to the original papers: Bessis and Pusterla [1967, 1968], Copley and Masson [1967], Copley et al. [1968], Basdevant et al. [1968, 1969], Basdevant and Lee [1969a], and Bessis and Turchetti [1972].

4.3 Lattice-Cutoff $\lambda\phi_n^4$ Euclidean Field Theory, or the Continuous-Spin Ising Model

In this section we address some of the ways in which Padé approximants can be used to study the deeper questions of the existence of a field theory and its relation to the renormalized perturbation theory. The approach here is to relate field theory to a corresponding problem in statistical mechanics. An important step is Nelson's theorem [1973], which showed that boson field theory could be completely studied in Euclidean space and that the corresponding Minkowski-space theory could always be constructed from the Euclidean theory. In Euclidean space the noncommuting operators of the usual field theory are replaced by correlated random fields. Once in Euclidean space, a lattice-type ultraviolet cutoff is used. The prototype Lagrangian, $\mathcal{L}$ [similar to (4.2.1), except with just one field component] and the action, $\mathcal{C}$, are related by

$$\mathcal{C}=\int dx\,\mathcal{L}(\phi(x))=\frac{1}{2}\int_{-\infty}^{+\infty}\cdots\int dx\left\{\left[\nabla\phi(x)\right]^2+m_0^2:\phi^2:+\tfrac{1}{12}\lambda_0:\phi^4:\right\},$$

$$(3.1)$$

where m_0 is the bare mass, λ_0 is the bare coupling constant, and $:\phi^{2p}:$ is the normal-ordered product. For the present purpose, $:\phi^{2p}:$ can be thought of as an even monic polynomial in ϕ of degree $2p$ whose coefficients are numerical functions of m_0^2, the spatial dimension and the ultraviolet cutoff. These coefficients in a space of two or more dimensions become infinite when the cutoff is removed. The lattice-cutoff fundamental function is the partition function

$$Z(H) = M^{-1} \int_{-\infty}^{+\infty} \cdots \int \prod_{i=1}^{N} d\phi_i$$

$$\times \exp\left[-\frac{v}{2} \sum_i \left\{ \frac{2d}{q} \sum_{\{\delta\}} \frac{(\phi_i - \phi_{i+\delta})^2}{a^2} + m_0^2 :\phi_i^2: + \tfrac{1}{12}\lambda_0 :\phi_i^4: \right\} \right.$$

$$\left. + \sum_i H_i \phi_i \right], \qquad (3.2)$$

where $v \propto a^d$ is the specific volume per lattice site, d is the spatial dimension, a is the lattice spacing, q is the lattice coordination number, $\{\delta\}$ is the set of one-half the nearest-neighbor sites, H_i is the magnetic field at site i, and M is a normalizing constant.

The field theory is defined by the Schwinger functions

$$\langle \phi_a \phi_b \cdots \phi_v \rangle = \frac{\partial^n \ln Z}{\partial H_a \cdots \partial H_v} \bigg|_{H_i = 0}, \qquad (3.3)$$

scaled for appropriate (possibly lattice-spacing dependent) choices of m_0 and γ_0 in the limit as the lattice spacing goes to zero.

By a suitable rescaling of the ϕ field, we can reexpress Equation (3.2) as

$$Z(\tilde{H}) = M^{-1} \int_{-\infty}^{+\infty} \cdots \int \prod_i d\sigma_i \exp\left[\sum_i \left\{ K \sum_{\{\delta\}} \sigma_i \sigma_{i+\delta} - \tilde{\lambda}_0 \sigma_i^4 - \tilde{A}\sigma_i^2 + H_i \sigma_i \right\} \right], \qquad (3.4)$$

which is immediately of the form of the continuous-spin Ising model on a d-dimensional spatial lattice. The infinite-volume limit of the field theory is now just the long-studied thermodynamic limit in statistical mechanics. The limit as the lattice spacing goes to zero can be thought of as that in which the ratio of the limit between the lattice spacing and the other length in the problem (i.e. the correlation length) goes to zero. This situation corresponds precisely to the approach to the critical point, where the fluctuations are large and correlations are long-range. An example is the highest temperature at which a ferromagnet, a bar of iron for example, can support a sponta-neous magnetization (in the absence of a magnetic field). The illumination of the deep questions of field theory is seen in this light as equivalent to the study in statistical mechanics of the approach to the critical point. Baker and Kincaid [1979, 1981] have begun such a study, and obtained remarkable results by Padé methods and the methods of Section 1.3.

To see how this investigation is aided by Padé methods, we will look at a few of the applications to statistical mechanics of critical phenomena. This area is one of the best developed of the applications of Padé methods to practical problems. From Equation (3.4) we may define the usual free energy per spin as

$$f = -(\beta N)^{-1} \ln Z, \tag{3.5}$$

where $\beta = 1/kT$, k is Boltzmann's constant, and T is the temperature. The typical thermodynamic quantities of interest can be derived directly from (3.5). At the critical point the various thermodynamics quantities have branch-point singularities, whose nature is of considerable physical interest. One can develop power-series expansions in $K \propto 1/kT$, for example, of all of these quantities. It is then the idea to use Padé methods to find the critical point and to determine the nature of the branch-point singularities. Typical quantities of interest would be

$$C_H \propto \left. \frac{\partial^2 f}{\partial \beta^2} \right|_H \propto |K - K_c|^{-\alpha},$$

$$M \propto \left. \frac{\partial f}{\partial H} \right|_\beta \propto |K - K_c|^\beta, \tag{3.6}$$

$$\chi \propto \left. \frac{\partial^2 f}{\partial H^2} \right|_\beta \propto |K - K_c|^{-\gamma},$$

where C_H is the specific heat at constant magnetic field, M is the spontaneous magnetization, χ is the magnetic susceptibility, and K_c is the critical value of K. We remark that

$$M = 0 \quad \text{for} \quad K \leqslant K_c$$

and

$$M > 0 \quad \text{for} \quad K > K_c. \tag{3.7}$$

In the case of the spin-$\frac{1}{2}$ Ising model (limit as $\tilde{\lambda}_0 \to \infty$, with $\langle \sigma_i^2 \rangle = 1$ for $K = 0$ fixed) in two dimensions, a number of the critical properties are known exactly [McCoy and Wu, 1973]; for example, $\gamma = 1.75$, $\beta = \frac{1}{8}$, and $C_H \propto \ln|K - K_c|$, which corresponds to $\alpha = 0$. For higher dimensions ($d = 3$, 4, etc.), no exact solution is known. For statistical mechanics, $d = 3$ is of great physical interest, and for field theory $d = 4$ is.

Now, from Equation (3.6), the problem for analysis is to design efficient procedures to utilize series-expansion information under the hypothesis that the leading behavior near the critical temperature is

$$g(x) \sim A(x)\left(1 - \frac{x}{x_0}\right)^{-\phi} + B(x), \qquad (3.8)$$

with A and B regular near x_0. We seek estimates for x_0, ϕ, $A(x_0)$, etc. The procedures (i)–(vi) described in Part I, Section 2.2 are designed to give just these estimates, and estimates of the apparent errors are given by the methods of Part I, Section 2.3. Using these techniques, various workers [Domb, 1974] have generally been able to compute the critical-point locations to 4 or 5 decimals and the critical indices to 3 or 4.

Another way in which Padé techniques have been used for the spin-$\frac{1}{2}$ Ising model is in connection with the high-field expansions. If in Equation (3.4) we set all $H_i = H$, we can expand the free energy about an ordered (all spins parallel) state in powers of $\mu = e^{-2H}$. By use of a remarkable theorem [Lee and Yang, 1952] which states that for real K, $\tilde{\lambda}_0$, $\tilde{A}$ all the singularities in the the μ plane lie on the unit circle $|\mu| = 1$ and come from zeros of Z, we can show [Bessis et al., 1976], in terms of the variable

$$v = \frac{4\mu}{(1+\mu)^2}, \qquad (3.9)$$

that the magnetization is

$$M(v) = 2(\tanh H)\int_0^{(1+\cos\theta_0)/2} \frac{d\phi(x)}{1 - xv}, \qquad (3.10)$$

where, for $K < K_c$, $\theta_0 > 0$, defines a singularity free sector, $-\theta_0 < \arg\mu < \theta_0$, and $K \geqslant K_c$, $\theta_0 = 0$. Since $M(v)/\tanh H$ is a Stieltjes series, the results of Part I, Chapter 5 apply and lead to bounds on M, θ_0, etc.

In addition to the high-temperature and high-field expansions already discussed, the spin-$\frac{1}{2}$ Ising model also has low-temperature expansions. Here there is an added complication. Instead of just one nearest, physically interesting singularity, there is a ring of close singularities. By use of the very long series available and the G^3J approximants of (1.3.9) which result from the solution of a linear, first-order, inhomogeneous integral equation, even this complicated structure has been successfully approximated [Hunter and Baker, 1979].

Of these various methods, the high-temperature series (K expansion) has been applied to the analysis of what is called the dimensionless, renormalized coupling constant

$$\lambda \propto \frac{\partial^2\chi/\partial H^2}{\chi^2\xi^d}, \qquad (3.11)$$

where ξ is the correlation length in units of the lattice spacing. Strong numerical evidence was obtained that the field-theory results are correct in $d=1$, 2, and 3 dimensions and are given by the asymptotic series in λ_0. These results are in accord with and extend the rigorous results of the constructive field theorists for very small λ_0. They also agree with the Padé–Borel summation (Section 1.2) of the λ_0 series [Baker et al., 1978]. In four dimensions the numerical evidence points to the idea that $\lambda_0 : \phi^4$: theory is trivial (i.e., no scattering for any $0 \leqslant \lambda_0 \leqslant \infty$), which leaves as an unanswered deep question the reason for the excellent results described in Section 4.2, based on pre-renormalized field theory. In the case of three dimensions an interesting feature was found. The limits of the lattice spacing going to zero (ultraviolet cutoff) and of the bare coupling constant λ_0 going to infinity do not commute. Consequently, the spin-$\frac{1}{2}$ Ising model is not described by the strong-coupling limit of the field theory as it was in one and two dimensions. Thus Padé analysis has found in this well-studied physical problem a richer structure than had been dreamed of.

Bibliography

References

Abd-Elall, L. F., Delves, L. M., and Reid, J. K. (1970), "A numerical method for locating the zeroes and poles of a meromorphic function," in P. Rabinowitz, (ed.), *Numerical Methods for Nonlinear Algebraic Equations*, Gordon and Breach, London, pp. 47–59.

Abramowitz, M. and Stegun, I. (1964), *Handbook of Mathematical Functions*, Dover.

Aitken, A. C. (1926), "On Bernoulli's numerical solution of algebraic equations," *Proc. Roy. Soc. Edin.*, **46**, 289–305.

Aitken, A. C. (1964), *Determinants and Matrices*, Oliver and Boyd.

Akhiezer, N. I. (1965), *The Classical Moment Problem*, Oliver and Boyd.

Alabiso, C. and Butera, P. (1975), "*N*-variable rational approximants and method of moments," *J. Math. Phys.*, **16**, 840–841.

Alabiso, C., Butera, P., and Prosperi, G. M. (1970), "Resolvent operator, Padé approximation and bound states in potential scattering," *Lett. Nuovo. Cim.*, **3**, 831–839.

Alabiso, C., Butera, P., and Prosperi, G. M. (1971), "Variational principles and Padé approximants. Bound states in potential theory," *Nucl. Phys. B*, **31**, 141–162.

Alabiso, C., Butera, P., and Prosperi, G. M. (1972a), "Variational principles and Padé approximants. Resonances and phase shifts in potential scattering," *Nucl. Phys. B*, **42**, 493–517.

Alabiso, C., Butera, P. and Prosperi, G. M. (1972b), "Variational Padé solution of the Bethe–Salpeter equation," *Nucl. Phys. B*, **46**, 593–614.

Alder, K., Trautmann, D., and Viollier, R. D. (1973), "Anwendung der Padé Approximation in der Kernphysik," *Z. für Naturforschung*, **28**, 321–331.

Allen, G. D., Chui, C. K., Madych, W. R., Narcowich, F. J., and Smith, P. W. (1974), "Padé approximation and Gaussian quadrature," *Bull. Austr. Math. Soc.*, **10**, 263–270.

Allen, G. D., Chui, C. K., Madych, W. R., Narcowich, F. J., and Smith, P. W. (1975), "Padé approximants of Stieltjes series," *J. Approx. Theory*, **14**, 302–316.

Allen, G. D. and Narcowich, F. J. (1975), "On representation and approximation of a class of operator-valued analytic functions," *Bull. Am. Math. Soc.*, **81**, 410–412.

Amos, A. T. (1978), "Padé approximants and Rayleigh Schrödinger perturbation theory," *J. Phys. B*, **11**, 2053–2060.

Anderson, G. (1740), Letter from Anderson in Leyden to Jones, in S. J. Rigaud (ed.), *Correspondence of Scientific Men of the Seventeenth Century*, George Olms, Hildesheim (1965), Vol. I, p. 361.

Argyris, J. H., Vaz, L. E., and William, K. J. (1977), "Higher order methods for transient diffusion analysis," *Computer Methods in Appl. Mech. and Engineering*, **12**, 243–278.

Arms, R. J. and Edrei, A. (1970), "The Padé tables and continued fractions generated by totally positive sequences," in *Mathematical Essays*, Ohio University Press, Athens, Ohio, pp. 1–21.

Axelsson, O. (1969), "A class of A-stable methods," *B.I.T.*, **9**, 185–199.

Axelsson, O. (1972), "A note on a class of strongly A-stable methods." *B.I.T.*, **12**, 1–4.

Baker, G. A., Jr. (1960), "An implicit, numerical method for solving the *n*-dimensional heat equation," *Quarterly of Appl. Math.*, **17**, 440–443.

Baker, G. A., Jr. (1961) "Application of the Padé approximant method to the investigation of some magnetic properties of the Ising model," *Phys. Rev.*, **124**, 768–774.

Baker, G. A., Jr. (1965), "The theory and application of the Padé approximant method," *Adv. in Theoretical Phys.*, **1**, 1–58.

Baker, G. A., Jr. (1969), "Best error bounds for Padé approximants to convergent series of Stieltjes," *J. Math. Phys.*, **10**, 814–820.

Baker, G. A., Jr. (1970), "The Padé approximant and some related generalizations", in G. A. Baker, Jr., and J. L. Gammel (eds.), *The Padé Approximant in Theoretical Physics*, Academic Press, New York, pp. 1–39.

Baker, G. A., Jr. (1972), "Converging bounds for the free energy in certain statistical mechanical problems," *J. Math. Phys.*, **13**, 1862–1864.

Baker, G. A., Jr. (1973a), "Recursive calculation of Padé approximants," in P. R. Graves-Morris (ed.), *Padé Approximants and Their Applications*, Academic Press.

Baker, G. A., Jr. (1973b), "Existence and Convergence of Subsequences of Padé approximants," *J. Math. Anal. Appl.*, **43**, 498–528.

Baker, G. A., Jr. (1974), "Certain invariance properties of the Padé approximant," *Rocky Mtn. J. Math.*, **4**, 141–149.

Baker, G. A., Jr. (1975), "Convergence of Padé approximants using the solution of linear functional equations," *J. Math. Phys.*, **16**, 813–822.

Baker, G. A., Jr. (1976), "A theorem on convergence of Padé approximants," *Studies in Applied Math.*, **55**, 107–117.

Baker, G. A., Jr. (1977), "The application of Padé approximants to critical phenomena," in E. B. Saff and R. H. Varga (eds.), *Padé and Rational Approximation*, Academic Press, New York, pp. 323–337.

Baker, G. A., Jr. and Chisholm, J. S. R. (1966), "The validity of perturbation series with zero radius of convergence," *J. Math. Phys.*, **7**, 1900–1902.

Baker, G. A., Jr. and Gammel, J. L. (1961), "The Padé approximant," *J. Math. Anal. Appl.*, **2**, 21–30.

Baker, G. A., Jr. and Gammel, J. L. (1971), "Application of the principle of the minimum maximum modulus to generalized moment problems and some remarks on quantum field theory," *J. Math. Anal. Appl.*, **33**, 197–211.

Baker, G. A., Jr., Gammel, J. L., and Wills, J. G. (1961), "An investigation of the applicability of the Padé approximant method," *J. Math. Anal. Appl.*, **2**, 405–418.

Baker, G. A., Jr., Gilbert, H. E., Eve, J., and Rushbrooke, G. S. (1967), "High temperature expansions for the spin-$\frac{1}{2}$ Heisenberg model," *Phys. Rev.*, **164**, 800–817.

Baker, G. A., Jr. and Graves-Morris, P. R. (1977), "Convergence of rows of the Padé table," *J. Math. Anal. Appl.*, **57**, 323–339.

Baker, G. A., Jr. and Gubernatis, J. E. (1981), "An asymptotic, Padé approximant method for

Legendre series," in P. L. Butzer (ed.), *Proc. 1979 Int. Christoffel Symposium*, Birkhäuser Verlag, to appear.

Baker, G. A., Jr. and Hunter, D. L. (1973), "Methods of series analysis II: Generalized and extended methods with application to the Ising model," *Phys. Rev. B* **7**, 3377–3392.

Baker, G. A., Jr. and Kincaid, J. M. (1979), "The continuous spin Ising model and $\lambda : \phi^4 :_d$ field theory," *Phys. Rev. Lett.*, **42**, 1431–1434; *Errata*, **44**, 434 (1980).

Baker, G. A., Jr. and Kincaid, J. M. (1980), "The continuous-spin Ising model, $g_0 : \phi^4 :_d$ field theory and the renormalization group," J. Stat. Phys., in press.

Baker, G. A., Jr. and Moussa, P. (1978), "Relation in the Ising model of the Lee–Yang branch point and critical behaviour," *J. Appl. Phys.*, **49**, 1360–1362.

Baker, G. A., Jr., Nickel, B. G., Green, M. S., and Meiron, D. I. (1976), "Ising model critical indices in three dimensions from the Callan–Symanzik equation," *Phys. Rev. Lett.*, **36**, 1351–1354.

Baker, G. A., Jr. and Oliphant, T. A. (1960), "An implicit, numerical method for solving the two-dimensional heat equation," *Quarterly of Appl. Math.*, **17**, 361–373.

Bareiss, E. H. (1969), "Numerical solution of linear equations with Toeplitz matrices," *Num. Math.*, **13**, 404–424.

Barnsley, M. F. (1973), "The bounding properties of the multipoint Padé approximant to a series of Stieltjes on the real line," *J. Math. Phys.*, **14**, 299–313.

Barnsley, M. F. (1976), "Padé approximant bounds for the difference of two series of Stieltjes," *J. Math. Phys.*, **17**, 559–565.

Barnsley, M. F. and Baker, G. A., Jr. (1976), "Bivariational bounds in a complex Hilbert space, and correction terms," *J. Math. Phys.*, **17**, 1019–1027.

Barnsley, M. F. and Bessis, D. (1979), "Constructive methods based on analytic characterizations and their applications to non-linear elliptic and parabolic differential equations," *J. Math. Phys.*, **20**, 1135–1145.

Barnsley, M. F., Bessis, D., and Moussa, P. (1979), "The Diophantine moment problem and the analytic structure in the activity of the ferromagnetic Ising model," *J. Math. Phys.*, **20**, 535–546.

Barnsley, M. F. and Robinson, P. D. (1974a), "Padé approximant bounds and approximate solutions for Kirkwood Riseman integral equations," *J. Inst. Math. Appl.*, **14**, 251–285.

Barnsley, M. F. and Robinson, P. D. (1974b), "Bivariational Bounds," *Proc. Roy. Soc. London A* **338**, 527–533.

Barnsley, M. F. and Robinson, P. D. (1974c), "Dual variational principles and Padé-type approximants," *J. Inst. Math. Appl.*, **14**, 229–249.

Barnsley, M. F. and Robinson, P. D. (1978), "Rational approximant bounds for a class of two variable Stieltjes functions," *SIAM J. Math. Anal.*, **9**, 272–290.

Basdevant, J.-L. (1969), "Padé approximants," in M. Nikolic (ed.), *Methods in Subnuclear Physics*, Gordon and Breach, New York, pp. 1–40.

Basdevant, J.-L. (1972), "Padé approximants in strong interaction physics," in D. Bessis (ed.), *Cargèse Lectures in Physics*, vol. 5, Gordon and Breach, New York, pp. 431–459.

Basdevant, J.-L. (1973), "Strong interaction physics and the Padé approximation in quantum field theory," in P. R. Graves-Morris (ed.), *Padé Approximants*, Inst. of Physics, Bristol, pp. 77–100.

Basdevant, J.-L., Bessis, D., and Zinn-Justin, J. (1968), "Padé approximants in strong interactions. Two body pion and Kaon systems," *Phys. Lett.*, **27B**, 230–233.

Basdevant, J.-L., Bessis, D., and Zinn-Justin, J. (1969), "Padé approximants in strong interactions. Two body pion and Kaon systems," *Nuovo Cim.*, **60A**, 185–238.

Basdevant, J.-L. and Lee, B. W. (1969a), "Padé approximants in the σ-model with unitary $\pi-\pi$ amplitudes with the current algebra constraints," *Phys. Lett.*, **29B**, 437–441.

Basdevant, J.-L. and Lee, B. W. (1969b), "Padé approximants and bound states: exponential potential," *Nucl. Phys. B*, **13**, 182–188.

Bauer, F. L. (1959), "The quotient-difference and epsilon algorithm," in R. Langer (ed.), *Numerical Approximation*, Univ. of Wisconsin Press, Madison, pp. 361–370.

Bauer, F. L., Rutishauser, H., and Stiefel, E. (1963), "New aspects in numerical quadrature," in N. C. Metropolis, A. H. Traub, J. Todd, and C. B. Tompkins (eds.), *Proc. of 15th Symposium in Applied Mathematics*, Amer. Math. Soc. Press, Vol. 15, pp. 199–218.

Bauhoff, W. (1977), "Padé approximants in the Wick–Cutkosky model," *J. Phys. A*, **10**, 129–134.

Beardon, A. F. (1968a), "The convergence of Padé approximants," *J. Math. Anal. Appl.*, **21**, 344–346.

Beardon, A. F. (1968b), "On the location of poles of Padé approximants," *J. Math. Anal. Appl.*, **21**, 469–474.

Bell, G. I. and Glasstone, S. (1970), *Nuclear Reactor Theory*, van Nostrand-Reinhold, New York.

Bender, C. and Wu, T. T. (1968), "Analytic structure of energy levels in a field theory model," *Phys. Rev. Lett.*, **21**, 406–409.

Benofy, L. P. and Gammel, J. L. (1977), "Variational principles and matrix Padé approximants," in E. B. Saff and R. H. Varga (ed.), *Padé and Rational Approximation*, Academic Press, New York, 339–356.

Benofy, L. P., Gammel, J. L., and Mery, P. (1976), "The off-shell momentum as a variational parameter in calculations of matrix Padé approximants in potential scattering," *Phys. Rev. D*, **13**, 3111–3114.

Bernstein, S. (1928), "Sur les fonctions absolument monotones," *Acta Math.*, **52**, 1–66.

Bessis, D., ed. (1972), *Cargèse Lectures in Physics*, Vol. 5, Gordon and Breach, New York.

Bessis, D. (1973), P. R. Graves-Morris (ed.), *Topics in the Theory of Padé Approximants*, Inst. of Phys., Bristol, 19–44.

Bessis, D. (1976), "Construction of variational bounds for the N-body eigenstate problem by the method of Padé approximants," in *Padé Approximants Method and its Application to Mechanics*, H. Cabannes (ed.), Springer Lecture Notes in Phys., No. 37, Berlin, 17–31.

Bessis, D. (1979), "A new method for the combinatorics of the topological expansion," *Comm. Math. Phys.*, **69**, 147–163.

Bessis, D., Drouffe, J. M., and Moussa, P. (1976), "Positivity constraints for the Ising ferromagnetic model," *J. Phys. A*, **9**, 2105–2124.

Bessis, D., Epele, L., and Villani, M. (1974), "Summation of regularized perturbation expansions for singular interactions," *J. Math. Phys.*, **15**, 2071–2078.

Bessis, D., Mery, P., and Turchetti, G. (1974), "Angular momentum analysis of four nucleon Green's function," *Phys. Rev. D*, **10**, 1992–2009.

Bessis, D., Mery, P., and Turchetti, G. (1977), "Variational bounds from matrix Padé approximants in potential scattering," *Phys. Rev. D*, **15**, 2345–2353.

Bessis, D., Moussa, P., and Villani, M. (1975), "Monotonic converging variational approximants to the functional integrals in quantum statistical mechanics," *J. Math. Phys.*, **16**, 2318–2325.

Bessis, D. and Pusterla, M. (1967), "Unitary Padé approximants for the S-matrix in strong coupling field theory and application to the calculation of ρ and f^0 Regge trajectories," *Phys. Lett.*, **25B**, 279–281.

Bessis, D. and Pusterla, M. (1968), "Unitary Padé approximants in strong coupling field theory and application to the ρ and f^0 meson trajectories," *Nuovo Cim.*, **54A**, 243–294.

Bessis, D. and Turchetti, G. (1972), "Renormalization of the σ-model through Ward identities," in D. Bessis (ed.), *Cargèse Lectures in Physics*, Vol. 5, Gordon and Breach, New York, 119–178.

Bessis, D. and Turchetti, G. (1977), "Variational matrix Padé approximants in potential scattering and low energy Lagrangian field theory," *Nucl. Phys. B*, **123**, 173–188.

Bessis, D. and Villani, M. (1975), "Perturbative-variational approximations to the spectral properties of semi-bounded Hilbert space operators, based on the moment problem with finite or diverging moments. Application to quantum mechanical systems," *J. Math. Phys.*, **16**, 462–474.

Biswas, S. N. and Vidhani, T. (1973), "Solution of Schrödinger equation with continued fractions," *J. Phys. A*, **6**, 468–477.

Blanch, G. (1964), "The numerical evaluation of continued fractions," *SIAM Rev.*, **6**, 383–421.

Bombelli, R. (1972), *L'Algebra*, Venezia.

Boas, R. P. (1954), *Entire Functions*, Academic Press, New York.

Brent, R. P., Gustavson, F. G. and Yun, D. Y. Y. (1980), "Fast solution of Toeplitz systems of equations and computation of Padé approximants", *J. Algorithms*, **1**, 259–295.

Brezin, E., Le Guillou, J. C., and Zinn-Justin, J. (1977), "Perturbation theory at large order. I. The ϕ^{2^N} interaction," *Phys. Rev. D*, **15**, 1544–1558.

Brezinski, C. (1971), "Accélération de suites à convergence logarithmique," *Comptes Rendus*, **273A**, 727–730.

Brezinski, C. (1972), "Conditions d'application et de convergence de procédés d'extrapolation," *Num. Math.*, **20**, 64–79.

Brezinski, C. (1974), "Review of acceleration methods," *Rendiconti di Mat.*, **7**, 303–316.

Brezinski, C. (1976), "Computation of Padé approximants and continued fractions," *J. Comp. Appl. Math.*, **2**, 113–123.

Brezinski, C. (1977), *Accélération de la Convergence en Analyse Numérique*, Springer Lecture Notes in Mathematics, No. 584, Berlin.

Brezinski, C. (1978a), "Convergence acceleration of some sequences by the ε-algorithm," *Num. Math.*, **29**, 173–177.

Brezinski, C. (1978b), "Padé type approximants for double power series," *J. Indian Math. Soc.*, **42**, 267–282.

Brezinski, C. (1979), "Rational approximation to formal power series," *J. Approx. Theory*, **25**, 295–317.

Brezinski, C. (1980), "A general extrapolation algorithm," *Num. Math.*, **35**, 175–187.

Brezinski, C. and Rieu, A. C. (1974), "Solution of systems of equations using the ε-algorithm, and application to boundary value problems," *Math. Comp.*, **28**, 731–734.

Bulirsch, R. and Stoer, J. (1964), "Fehler abschätzungen und Extrapolation mit rationalen Functionen bei Verfahren vom Richardson-Typus," *Num. Math.*, **6**, 413–427.

Bultheel, A. (1979), "Recursive algorithms for the Padé table: two approaches," in L. Wuytack (ed.), *Padé Approximation and Its Applications*, Springer Lecture Notes in Mathematics, No. 765, pp. 211–230.

Bultheel, A. (1980a), "Recursive algorithms for the matrix Padé problem," *Math. Comp.*, **35**, 875–892.

Bultheel, A. (1980b), "Division algorithms for continued fractions and the Padé table," *J. Comp. Appld. Math.*, **6**, 259–266.

Bultheel, A. and Wuytack, L., (1981) "Stability of numerical methods for computing Padé approximants." in *Approximation Theory* III, ed. E. W. Cheney, Academic Press, New York.

Bus, J. C. P. and Dekker, T. J. (1975), "Two efficient algorithms with guaranteed convergence for finding a zero of a function," *ACM Trans. on Math. Software*, **1**, 330–345.

Butcher, J. C. (1963), "Coefficients for the study of Runge-Kutta integration processes," *J. Australian Math. Soc.*, **3**, 185–201.

Butcher, J. C. (1964), "Implicit Runge-Kutta Processes," *Math. Comp.*, **18**, 50–64.

Carleman, T. (1926), *Les Fonctions Quasi-Analytiques*, Gauthier Villars, Paris; translated by J. L. Gammel, Los Alamos Tech. report 4702, (1971).

Carroll, A., Baker, G. A., Jr., and Gammel, J. L. (1977), "A comment on a paper of Hamer applying Padé approximants to a $1+1$ dimensional lattice model of Q.C.D., *Nucl. Phys. B*, **129**, 361–364.

Carslaw, H. S. and Jaeger, J. C. (1959), *Conduction of Heat in Solids*, Oxford University Press.

Cartan, H. (1928), "Sur les systèmes de fonctions holomorphes à variétés linéaires et leurs applications," *Ann. Sci. Ecole. Norm. Sup.* (*3*), **45**, 255–346.

Caser, S., Piquet, C., and Vermeulen, J. L. (1969), "Padé approximants for a Yukawa potential," *Nucl. Phys. B*, **14**, 119–132.

Cavendish, J. C., Culham, W. E., and Varga, R. S. (1972), "A comparison of Crank–Nicholson and Chebychev rational methods for numerically solving linear parabolic equations," *J. Comp. Phys.*, **10**, 354–368.

Cauchy, M. A.-L. (1821), *Cours d'analyse* de l'Ecole Royale Polytechnique; première partie, L'Imprimerie Royale, Paris.

Char, B. W. (1980), "On Stieltjes continued fraction for the Gamma Function," *Math. Comp.*, **34**, 547–551.

Cheney, E. W. (1966), *Introduction to Approximation Theory*, McGraw-Hill.

Chipman, F. H. (1971), "A-stable Runge-Kutta processes," *B.I.T.*, **11**, 384–388.

Chisholm, J. S. R. (1963), "Solution of linear integral equations using Padé approximants," *J. Math. Phys.*, **4**, 1506–1510.

Chisholm, J. S. R. (1966), "Approximation by sequences of Padé approximants in regions of meromorphy," *J. Math. Phys.*, **7**, 39–46.

Chisholm, J. S. R. (1973), "Rational approximants defined from double power series," *Math. Comp.*, **27**, 841–848.

Chisholm, J. S. R. (1977a), "N-variable rational approximants," in E. B. Saff and R. H. Varga (eds.), *Padé and Rational Approximation*, Academic Press, New York, 23–42.

Chisholm, J. S. R. (1977b), "Multivariate approximants with branch points. I. Diagonal approximants," *Proc. Roy. Soc. Lond. A*, **358**, 351–366.

Chisholm, J. S. R. (1978a), "Multivariate approximants with branch cuts" in *Multivariate Approximation*, ed. D. C. Handscomb, Academic Press, London.

Chisholm, J. S. R. (1978b), "Multivariate approximants with branch points. II. Off-diagonal approximants," *Proc. Roy. Soc. Lond. A*, **362**, 43–56.

Chisholm, J. S. R. and Common, A. K. (1980), "Generalizations of Padé approximation for Chebychev and Fourier series," in P. L. Butzer (ed.), *Proc. 1979 Int. Christoffel Symposium*, Birkhäuser Verlag.

Chisholm, J. S. R., Genz, A. C., and Rowlands, G. E. (1972), "Accelerated convergence of sequences of quadrature approximations," *J. Comp. Phys.*, **10**, 284–307.

Chisholm, J. S. R. and Graves-Morris, P. R. (1975), "Generalizations of the theorem of de Montessus to two-variable approximants," *Proc. Roy. Soc. Lond. A*, **342**, 341–372.

Chisholm, J. S. R. and Hughes Jones, R. (1975), "Relative scale covariance of N-variable approximants," *Proc. Roy. Soc. Lond. A*, **344**, 465–470.

Chisholm, J. S. R. and McEwan, J. (1974), "Rational approximants defined from power series in N-variables," *Proc. Roy. Soc. Lond. A*, **336**, 421–452.

Chisholm, J. S. R. and Roberts, D. E. (1976), "Rotationally covariant approximants derived from double power series," *Proc. Roy. Soc. Lond. A*, **351**, 585–591.

Christoffel, E. B. (1877), "Sur une classe particulière de fonctions entières et de fractions continues," *Ann. di Mat.*, **8**, 1–10.

Chui, C. K., Shisha, O., and Smith, P. W. (1974), "Padé approximants as limits of best rational approximants," *J. Approx. Theory*, **12**, 201–204.

Chui, C. K., Shisha, O., and Smith, P. W. (1975), "Best local approximation," *J. Approx. Theory*, **15**, 371–381.

Chui, C. K., Smith, P. W., and Ward, J. D. (1978), "Best L_2 local approximation," *J. Approx. Theory*, **22**, 254–261.

Claessens, G. (1976), "A new algorithm for osculatory rational interpolation," *Num. Math.*, **27**, 77–83.

Claessens, G. (1978a), "On the Newton-Padé approximation problem," *J. Approx. Theory*, **22**, 150–160.

Claessens, G. (1978b), "On the structure of the Newton-Padé table," *J. Approx. Theory*, **22**, 304–319.

Claessens, G. (1978c), "Generalized ε-algorithm for rational interpolation," *Num. Math.*, **29**, 227–231.

Claessens, G. and Wuytack, L. (1979), "On the computation of non-normal Padé approximants," *J. Comp. Appl. Math.*, **5**, 283–289.

Clenshaw, C. W. and Lord, K. (1974), "Rational approximations from Chebychev series," in B. K. P. Scaife (ed.), *Studies in Numerical Analysis*, Academic Press, London, 95–113.

Cody, W. J., Meinardus, G., and Varga, R. S. (1969), "Chebychev rational approximations to e^{-x} in $[0, \infty)$ and applications in heat conduction equations," *J. Approx. Theory*, **2**, 50–65.

Common, A. K. (1968), "Padé approximants and bounds to series of Stieltjes," *J. Math. Phys.*, **9**, 32–38.

Common, A. K. (1969a), "Properties of Legendre expansions related to Stieltjes series and applications to π–π scattering," *Nuovo Cim.*, **63A**, 863–891.

Common, A. K. (1969b), "Properties of generalisations to Padé approximants," *J. Math. Phys.*, **10**, 1875–1880.

Common, A. K. (1970), "Some consequences of relations of π–π partial wave amplitudes to Stieltjes series," *Nuovo Cim.*, **65A**, 581–596.

Common, A. K. (1979), "Calculation of Yukawa scattering amplitude and impact parameter amplitudes using Legendre–Padé approximants," *J. Phys. A*, **12**, 2563–2572.

Common, A. K. and Graves-Morris, P. R. (1974), "Some properties of Chisholm approximants," *J. Inst. Math. Appl.*, **13**, 229–232.

Common, A. K. and Stacey, T. W. (1979a), "The convergence of Legendre–Padé approximants to the Coulomb and other scattering amplitudes," *J. Phys. A*, **11**, 275–289.

Common, A. K. and Stacey, T. W. (1979b), "Legendre–Padé approximants and their application in potential scattering," *J. Phys. A*, **11**, 259–273.

Common, A. K. and Stacey, T. W. (1979c), "Convergent series of Legendre–Padé approximants to the real and imaginary parts of the scattering amplitudes," *J. Phys. A*, **12**, 1399–1417.

Conn, R. W. (1974), "Higher order variational principles and Padé approximants for linear functionals," *Nucl. Sci. Eng.*, **55**, 468–470.

Copley, L. A., Elias, D. K., and Masson, D. (1968), "Broken symmetry model of meson interactions," *Phys. Rev.*, **173**, 1552–1563.

Copley, L. A. and Masson, D. (1967), "Padé approximant calculation of π–π scattering," *Phys. Rev.*, **164**, 2059–2062.

Copson, E. T. (1948), *An Introduction to the Theory of a Complex Variable*, Oxford University Press.

Corcoran, C. T. and Langhoff, P. W. (1977), "Moment theory approximations for non-negative spectral densities," *J. Math. Phys.*, **18**, 651–657.

Cordellier, F. (1979a), "Sur la régularité des procédés δ^2 d'Aitken et W de Lubkin," in L. Wuytack (ed.), *Padé Approximation and Its Applications*, Springer Lecture Notes in Math., No. 765, Berlin.

Cordellier, F. (1979b), Demonstration algébrique de l'extension de l'identité de Wynn aux tables de Padé non-normales," in L. Wuytack (ed.), *Padé Approximation and Its Applications*, Springer Lecture Notes in Math., No. 765, Berlin.

Courant, R. and Hilbert, D. (1953), *Methods of Mathematical Physics*, Vol. I, Interscience.

Crank, J. (1975), *The Mathematics of Diffusion*, Oxford University Press.

Dekker, T. J. (1969), "Finding a zero by means of successive linear interpolation," in B. Dejon and P. Henrici (eds.), *Constructive Aspects of the Fundamental Theorem of Algebra*, Wiley, 37–48.

Delves, L. M. and Phillips, A. C. (1969), "Present status of the nuclear three body problem," *Rev. Mod. Phys.*, **41**, 497–530.

Dennis, J. J. and Wall, H. S. (1945), "The limit-circle case for a positive definite J-fraction," *Duke Math. J.*, **12**, 255–273.

Dienes, P. (1957), *The Taylor series*, Dover, New York.

Dijkstra, D. (1977), "A continued fraction expansion for a generalization of Dawson's integral," *Math. Comp.*, **31**, 503–510.

Dimock, J. (1974), "Asymptotic perturbation expansion in the $P(\phi)_2$ quantum field theory," *Comm. Math. Phys.*, **35**, 347–356.

Dirac, P. A. M. (1958), *Principles of Quantum Mechanics*, Oxford University Press.

Domb, C. (1974), "Ising model," in C. Domb and M. S. Green (eds.), *Phase Transitions and Critical Phenomena*, vol. 3, Academic Press, London, 357–485.

Domb, C. and Sykes, M. F. (1961), "Use of series expansions for the Ising model susceptibility and excluded value problem," *J. Math. Phys.*, **2**, 63–67.

Donnelly, J. D. P. (1966), "The Padé table," in D. C. Handscomb (ed.), *Methods of Numerical Approximation*, Pergamon Press, Oxford, 125–130.

Drew, D. and Murphy, J. A. (1977), "Branch points, M-fractions and rational approximants generated by linear equations," *J. Inst. Math. Appl.*, **19**, 169–185.

Dyson, F. J. (1952), "Divergence of Perturbation Theory in Quantum Electrodynamics," *Phys. Rev.* **85**, 631–632.

Eckmann, J.-P., Magnen, J., and Sénéor, R. (1974), "Decay properties and Borel summability for the Schwinger functions in $P(\phi)_2$ theories," *Comm. Math. Phys.*, **39**, 251–271.

Edrei, A. (1939), "Sur les determinants récurrents et les singularités d'une fonction données par son développement de Taylor," *Compositio Math.*, **7**, 20–88.

Edrei, A. (1953), "Proof of a conjective of Schoenberg on the generating function of a totally positive sequence," *Can. J. Math.*, **5**, 86–94.

Edrei, A. (1975a), "Convergence of complete Padé tables of trigonometric functions," *J. Approx. Theory*, **15**, 278–293.

Edrei, A. (1975b), "The Padé table of functions having a finite number of essential singularities," *Pacific J. Math.*, **56**, 429–453.

Ehle, B. L. (1968), "High order A-stable methods for numerical solution of systems of differential equations," *B.I.T.*, **8**, 276–278.

Ehle, B. L. (1973), "A-stable methods and Padé approximants to the exponential," *SIAM J. Math. Anal.*, **4**, 671–680.

Ehle, B. L. (1976), "On certain order constrained Chebychev rational approximations," *J. Approx. Theory*, **17**, 297–306.

Erdélyi, A. (1956), *Asymptotic Expansions*, Dover, New York.

Euler, L. (1737), "De fractionibus continuis," *Comm. Acad. Sci. Imper. Petropol.*, **9**, 98–137.

Fair, W. G. (1964), "Padé approximants to the solution of the Ricatti equation," *Math. Comp.*, **18**, 627–634.

Fairweather, G. (1971), "A survey of discrete Galerkin methods for parabolic equations in one space variable," *Math. Colloq. U.C.T.*, **7**, 43–77.

Feldman, J. S. and Osterwalder, K. (1976), "The Wightman axioms and the mass gap for weakly coupled $(\phi^4)_3$ quantum field theories," *Ann. Phys.*, **97**, 80–135.

Ferer, M., Moore, M. A., and Wortis, M. (1971), "Some critical properties of the nearest-neighbor, classical Heisenberg model for the f.c.c. lattice in finite field for temperatures greater than T_c," *Phys. Rev. B*, **4**, 3954–3963.

Ferrar, W. L. (1938), *Convergence*, Oxford University Press.

Fisher, M. E. (1977), "Series expansion approximants for singular functions of many variables," in Uzi Landman (ed.), *Statistical Mechanics and Statistical Methods in Theory and Application*, Plenum Press, pp. 3–31.

Fisher, M. E. and Kerr, R. M. (1977), "Partial differential approximants for multicritical singularities," *Phys. Rev. Lett.*, **39**, 667–670.

Fleischer, J. (1972), "Analytic continuation of scattering amplitudes and Padé approximants," *Nucl. Phys. B*, **37**, 59–76.

Fleischer, J. (1973a), "Nonlinear Padé approximants for Legendre series," *J. Math. Phys.*, **14**, 246–248.

Fleischer, J. (1973b), "Generalizations of Padé approximants," in P. R. Graves-Morris (ed.), *Padé Approximants*, Inst. of Phys., Bristol, pp. 126–131.

Fleischer, J., Gammel, J. L., and Menzel, M. T. (1973), "Matrix Padé approximants for the $^1S_0$ and $^3P_0$ partial waves in nucleon–nucleon scattering," *Phys. Rev. D*, **8**, 1545–1552.

Fleischer, J. and Tjon, J. A. (1975), "Bethe–Salpeter equation for $J=0$ nucleon–nucleon scattering with one boson exchange," *Nucl. Phys. B*, **84**, 375–396.

Fleischer, J. and Tjon, J. A. (1980), "Bethe–Salpeter equation for elastic nucleon–nucleon scattering," *Phys. Rev. D*, **21**, 87–94.

Fogli, G. L., Pellicoro, M. F., and Villani, M. (1971), "A summation method for a class of series with divergent terms," *Nuovo Cim.*, **6A**, 79–97.

Fogli, G. L., Pellicoro, M. F., and Villani, M. (1972), "An approach to the radiative corrections in Q.E.D. in the framework of the Padé method," *Nuovo Cim.*, **11A**, 153–177.

Ford, W. B. (1960), *Asymptotic Series and Divergent Series*, Chelsea, New York.

Frank, W. M., Land, D. J., and Spector, R. M. (1971), "Singular potentials," *Rev. Mod. Phys.*, **43**, 36–98.

Fratamico, G., Ortolani, F., and Turchetti, G. (1976) "Exact solutions from the variational [1/1] matrix Padé approximant in potential scattering," *Lett. Nuovo Cim.*, **17**, 582–584.

Frobenius, G. (1881), "Veber Relationen zwischen den Näherungsbrüchen von Potenzreihen," *J. für Reine und Angewandte Math.*, **90**, 1–17.

Gallucci, M. A. and Jones, W. B. (1976), "Rational approximations corresponding to Newton series (Newton–Padé approximants)," *J. Approx. Theory*, **17**, 366–392.

Gammel, J. L. (1973), "Review of two recent generalizations of the Padé approximant," in P. R. Graves-Morris (ed.), *Padé Approximants and Their Applications*, Academic Press, London, pp. 3–9.

Gammel, J. L. and McDonald, F. A. (1966), "Applications of the Padé approximant to scattering theory," *Phys. Rev.*, **142**, 1245–1254.

Gammel, J. L., Rousseau, C. C., and Saylor, D. P. (1967), "A Generalization of the Padé Approximant," *J. Math. Anal. Appl.*, **20**, 416–420.

Gargantini, J. and Henrici, P. (1967), "A continued fraction algorithm for the computation of higher transcendental functions in the complex plane," *Math. Comp.*, **21**, 18–29.

Garibotti, C. R. (1972), "Schwinger variational principle and Padé approximants," *Ann. Phys.*, **71**, 486–496.

Garibotti, C. R. and Grinstein, F. F. (1978a), "Summation of partial waves for long range potentials I," *J. Math. Phys.*, **19**, 821–829.

Garibotti, C. R. and Grinstein, F. F. (1978b), "Punctual Padé approximants as a regularization procedure for divergent and oscillatory partial wave expansion of the scattering amplitude," *J. Math. Phys.*, **19**, 2405–2409.

Garibotti, C. R. and Grinstein, F. F. (1979), "Summation of partial waves for long range potentials II," *J. Math. Phys.*, **20**, 141–147.

Garibotti, C. R., Pellicoro, M. F., and Villani, M. (1970), "Padé method in singular potentials," *Nuovo Cim.*, **66A**, 749–766.

Garibotti, C. R. and Villani, M. (1969a), "Continuation in the coupling constant for the total T and K matrices," *Nuovo Cim.*, **59A**, 107–123.

Garibotti, C. R. and Villani, M. (1969b), "Padé approximant and the Jost function," *Nuovo Cim.*, **61A**, 747–754.

Garnett, J. (1973), *Analytic Capacity and Measure*, Springer Lecture Notes in Mathematics, No. 297.

Garside, G. R., Jarratt, P., and Mack, C. (1968), "A new method for solving polynomial equations," *Comp. J.*, **11**, 87–90.

Gauss, C. F. (1813), "Disquisitiones generales circa seriam infinitam

$$1 + \frac{\alpha\beta}{1\cdot\gamma}x + \frac{\alpha(\alpha+1)\beta(\beta+1)}{1\cdot 2\gamma(\gamma+1)}x\cdot x + \cdots$$

Commentationes Societatis Regiae Scientiorum Goettingensis Recentiores, **2**.

Gauthier, P. M. (1977), "On the possibiltiy of rational approximation" in Padé and rational approximation, eds. E. B. Saff and R. S. Varga, Academic Press, New York, pp. 261–264.

Gautschi, W. (1967), "Computational aspects of three term recurrence relations," SIAM Review, **9**, 24–82.

Gear, C. W. (1971), Numerical Initial Value Problems in Ordinary Differential Equations, Prentice-Hall.

Geddes, K. O. (1979), "Symbolic computation of Padé approximants", A. C. M. Trans. Math. Software, **5**, 218–233.

Gekeler, E. (1972), "On the solution of systems of equations by the epsilon algorithm of Wynn," Math. Comp., **26**, 427–436.

Genz, A. C. (1972), "An adaptive multidimensional quadrature procedure," Comp. Phys. Comm., **4**, 11–15.

Genz, A. C. (1973), "The applications of the ε-algorithm to quadrature problems," in P. R. Graves-Morris (ed.), Padé Approximants, Inst. of Physics, Bristol, pp. 105–116.

Genz, A. C. (1974), "Some extrapolation methods for the numerical calculation of multidimensional integrals," in D. J. Evans (ed.), Software for Numerical Mathematics, Academic Press, New York, pp. 159–172.

Genz, A. C. (1977), "A non-linear method for the convergence of multidimensional sequences," J. Comp. Appl. Math., **3**, 181–184.

Germain-Bonne, B. (1979), "Ensembles des suites et de procédés liés pour l'accélération de la convergence," in L. Wuytack (ed.) Padé Approximation and Its Applications, Springer Lecture Notes in Mathematics, No. 765, pp. 116–134.

Gersten, A., Owen, D. A., Gammel, J. L., Mery, P., and Turchetti, G. (1976), "Matrix Padé approximants and the Bethe–Salpeter equation of the nucleon-nucleon interaction," Phys. Rev. D, **13**, 1140–1143.

Gilewicz, J. (1978), Approximants de Padé, Springer Lecture Notes in Mathematics, No. 667.

Giraud, B. G. (1978), "Lower bounds to bound state eigenvalues," Phys. Rev. C, **17**, 800–809.

Giraud, B. G., Khalil, A. B., and Moussa, P. (1976), "Padé approximants for distorted waves," Phys. Rev. C, **14**, 1679–1687.

Giraud, B. G. and Turchetti, G. (1978), "Rigorous lower bounds to the energy levels with finite summations," Lett. Nuovo Cim., **21**, 605–608.

Goldberger, M. L. and Watson, K. M. (1964), Collision Theory, Wiley, New York.

Goldhammer, P. and Feenberg, E. (1956), "Refinement of the Brillouin Wigner perturbation method," Phys. Rev., **101**, 1233–1234.

Gončar, A. A. and López, Guillermo L. (1978), Math. U.S.S.R. Sbornik, **34**, 449–459.

Gordon, R. G. (1968), "Error bounds in equilibrium statistical mechanics," J. Math. Phys., **9**, 655–663.

Götz, U., Rösel, F., Trautmann, D., and Jochin, H. (1976), "Influence of the Mott–Schwinger interaction on the elastic scattering of protons," Z. Phys. A, **278**, 139–143.

Graffi, S. and Grecchi, V. (1978), "Borel summability and indeterminacy of the Stieltjes moment problem: application to the anharmonic oscillators," *J. Math. Phys.*, **19**, 1002–1006.

Graffi, S., Grecchi, V., and Simon, B. (1970), "Borel summability: application to the anharmonic oscillator," *Phys. Lett.*, **32B**, 631–634.

Graffi, S., Grecchi, V., and Turchetti, G. (1971), "Summation methods for the perturbation series of the generalized harmonic oscillator," *Nuovo Cim.*, **4B**, 313–340.

Gragg, W. B. (1965), "On extrapolation algorithms for ordinary initial value problems," *SIAM J. Num. Anal.*, **2**, 384–403.

Gragg, W. B. (1968), "Truncation error bounds for g-fractions," *Num. Math.*, **11**, 370–379.

Gragg, W. B. (1970), "Truncation error bounds for π-fractions," *Bull. Am. Math. Soc.*, **76**, 1091–1094.

Gragg, W. B. (1972), "The Padé table and its relation to certain algorithms of numerical analysis," *SIAM Review*, **14**, 1–62.

Gragg, W. B. (1974), "Matrix interpretations and applications of the continued fraction algorithm," *Rocky Mtn. J. Math.*, **4**, 213–225.

Gragg, W. B. (1977), "Laurent, Fourier and Chebychev Padé tables," in E. B. Saff and R. H. Varga (eds.), *Padé and Rational Approximation*, pp. 61–70.

Gragg, W. B. and Johnson, G. D. (1974), "The Laurent Padé Table," *Info. Proc. 74*, North-Holland, Amsterdam, **3**, 632–637.

Graves-Morris, P. R. (1973), "Padé approximants and potential scattering," in P. R. Graves-Morris (ed.), *Padé Approximants*, Inst. of Phys., Bristol, pp. 64–76.

Graves-Morris, P. R. (1975), "Convergence of rows of the Padé table," in H. Cabannes (ed.), *Padé Approximation and Its Application to Mechanics*, Springer Lecture Notes in Physics, No. 47, pp. 55–68.

Graves-Morris, P. R. (1977), "Generalisations of the theorem of de Montessus using Canterbury approximants," in E. B. Saff and R. S. Varga (eds.), *Padé and Rational Approximation*, Academic Press, New York, pp. 73–82.

Graves-Morris, P. R. (1978a), "Padé approximants for integral equations?," *J. Inst. Math. Appl.*, **21**, 375–378.

Graves-Morris, P. R. (1978b), "Applications of matrix Padé approximants in potential theory," *Ann. Phys.*, **114**, 290–295.

Graves-Morris, P. R. (1979), "The numerical calculation of Padé approximants," in L. Wuytack (ed.), *Padé Approximation and Its Applications*, Springer Lecture Notes in Mathematics, No. 765, pp. 231–245.

Graves-Morris, P. R. (1980a), "Practical, reliable, rational interpolation," *J. Inst. Math. Appl.*, **25**, 267–286.

Graves-Morris, P. R. (1980b), "A generalized Q.D. algorithm," *J. Comp. Appl. Math.*, **6**, 247–249.

Graves-Morris, P. R. (1981), "The Convergence of Ray Sequences of Padé Approximants of Stieltjes Series" (preprint).

Graves-Morris, P. R. and Hopkins, T. R. (1981), "Reliable rational interpolation," *Num. Math.*, **36**, 111–128.

Graves-Morris, P. R., Hughes Jones, R., and Makinson, G. J. (1974), "The calculation of some rational approximants in two variables," *J. Inst. Math. Appl.*, **13**, 311–320.

Graves-Morris, P. R. and Rennison, J. F. (1974), "Analyticity in the coupling strength," *J. Math. Phys.*, **15**, 230–233.

Graves-Morris, P. R. and Roberts, D. E. (1975), "Calculation of Canterbury approximants," *Comp. Phys. Comm.*, **10**, 234–244; *Erratum*, (1977), **13**, 72.

Graves-Morris, P. R. and Samwell, C. J. (1975), "Canterbury approximants in potential scattering," *J. Phys. G.*, **1**, 805–814.

Gray, H. L., Atchison, T. A., and McWilliams, G. V. (1971), "Higher order G-transformations," *SIAM J. Num. Anal.*, **8**, 365–381.

Grenander, U. and Szegö, G. (1958), *Toeplitz Forms and Their Applications*, Univ. of California Press, Berkeley.

Grinstein, F. F. (1980), "Summation of partial wave expansions in the scattering by short range potentials," *J. Math. Phys.*, **21**, 112–119.

Guttmann, A. J. (1969), "Determination of critical behaviour in lattice statistics from series expansions III," *J. Phys. C*, **2**, 1900–1907.

Guttman, A. J. (1975a), "On the recurrence relation method of series analysis," *J. Phys. A*, **8**, 1081–1088.

Guttmann, A. J. (1975b), "Derivation of 'mimic expansions' from regular perturbation expansions in fluid mechanics," *J. Inst. Math. Appl.*, **15**, 307–315.

Guttmann, A. J. and Joyce, G. S. (1972), "On a new method of series analysis in lattice statistics," *J. Phys. A*, **5**, L81–L84.

Hadamard, J. (1892), "Essai sur l'étude des fonctions données par leur developpement de Taylor," $2^{\text{ième}}$ partie, *J. de Math.*, **4**, 101–186.

Hadamard, J. ("1968"), *Oeuvres*, Vol. 2, CNRS, Paris, pp. 24–60.

Hamburger, H. (1920), "Ueber eine Erweiterung des Stieltjes'schen Momenten problems," *Math. Annalen*, **81**, 235–319.

Hamburger, H. (1921), "Ueber eine Erweiterung des Stieltjes'schen Momentem problems," *Math. Annalen*, **82**, 120–164; 168–187.

Hardy, G. H. (1956), *Divergent Series*, Oxford University Press.

Haymaker, R. W. (1968), "Application of analyticity properties to the numerical solution of the Bethe–Salpeter equation," *Phys. Rev.*, **165**, 1790–1802.

Haymaker, R. W. and Schlessinger, L. (1970), "Padé approximants as a computational tool for solving the Schrödinger and Bethe–Salpeter equations," in G. A. Baker, Jr. and J. L. Gammel (eds.), *The Padé Approximant in Theoretical Physics*, Academic Press, New York, pp. 257–288.

Henrici, P. (1958), "The quotient difference algorithm," *Nat. Bur. Stand.—Appl. Math Series*, **49**, 23–46.

Henrici, P. (1964), *Elements of Numerical Analysis*, Wiley.

Henrici, P. (1974), *Applied and Computational Complex Analysis*, Vol. I, Wiley.

Henrici, P. and Pfluger, P. (1966), "Truncation error estimates for Stieltjes fractions," *Num. Math.*, **9**, 120–138.

Hildebrand, F. B. (1956), *"Introduction to Numerical Analysis,"* McGraw-Hill.

Hille, E. (1959), *Analytic Function Theory*, Vol. I, Ginn and Co.

Hille, E. (1962), *Analytic Function Theory*, Vol. II, Ginn and Co.

Hillion, P. (1977a), "Remarks on rational approximation of multiple power series," *J. Inst. Math. Appl.*, **19**, 281–293.

Hillion, P. (1977b), "Approximating functions with a given singularity," *J. Math. Phys.*, **18**, 465–470.

Hofman, H. M., Starkand, Y., and Kirson, M. W. (1976), *Nucl. Phys. A*, **266**, 138–162.

Holdeman, J. T., Jr. (1969), "A method for the approximation of functions defined by formal series expansion in orthogonal polynomials," *Math. Comp.*, **23**, 275–287.

Householder, A. S. (1970), *The Numerical Treatment of a Single Non-linear Equation*, McGraw-Hill, New York.

Householder, A. S. (1971), "The Padé table, the Frobenius identities and the q.d. algorithm," *Lin. Algebra and Its Appl.*, **4**, 161–174.

Householder, A. S. and Stewart, G. W., III (1969), "Bigradients, Hankel determinants and the Padé table," in B. Dejon and P. Henrici (eds.), *Constructive Aspects of the Fundamental Theorem of Algebra*, Academic Press, New York, pp. 131–150.

Hughes Jones, R. (1976), "General rational approximants in *N*-variables," *J. Approx. Theory*, **16**, 201–233.

Hughes Jones, R. and Makinson, G. J. (1974), "The generation of Chisholm rational approximants to power series in two variables," *J. Inst. Math. Appl.*, **13**, 299–310.

Hunter, D. L. and Baker, G. A., Jr. (1973), "Methods of series analysis I: comparison of methods used in critical phenomena," *Phys. Rev. B*, **7**, 3346–3376.

Hunter, D. L. and Baker, G. A., Jr. (1979), "Methods of series analysis III: integral approximant methods," *Phys. Rev. B*, **19**, 3808–3821.

Ince, E. L. (1944), *Ordinary Differential Equations*, Dover Publications, New York.

Isaacson, E. and Keller, H. B. (1966), *Analysis of Numerical Methods*, Wiley.

Isenberg, C. (1963), "Moment calculations in lattice dynamics I., F.C.C. lattice with nearest neighbor interactions," *Phys. Rev.*, **132**, 2427–2433.

Iserles, A. (1979), "A note on Padé approximations and generalized hypergeometric functions," B.I.T., **19**, 543–545.

Jacobi, C. G. J. (1846), "Über die Darstellung einer Reihe gegebner Werthe durch eine gebrochne rationale Function," *J. für Reine Angewandte Math.*, **30**, 127–156.

Jarratt, P. (1970), "A review of methods for solving non-linear algebraic equations in one variable," in P. Rabinowitz (ed.), *Numerical Methods for Non-linear Algebraic Equations*, Gordon and Breach, London, 1–26.

Jones, W. B. (1977), "Multiple point Padé tables," in E. B. Saff and R. H. Varga (eds.), *Padé and Rational Approximation*, Academic Press, New York, pp. 163–172.

Jones, W. B. and Thron, W. J. (1966), "Further properties of *T*-fractions," *Math. Annalen*, **166**, 106–118.

Jones, W. B. and Thron, W. J. (1971), "A posteriori bounds for the truncation error of continued fractions," *SIAM J. Num. Anal.*, **8**, 693–705.

Jones, W. B. and Thron, W. J. (1974), "Numerical stability in evaluating continued fractions," *Math. Comp.*, **28**, 795–810.

Jones, W. B. and Thron, W. J. (1975), "On the convergence of Padé approximants," *SIAM J. Math. Anal.*, **6**, 9–16.

Jones, W. B. and Thron, W. J. (1977), "Two point Padé tables and *T*-fractions," *Bull. Am. Math. Soc.*, **83**, 388–390.

Jones, W. B. and Thron, W. J. (1979), "Sequences of meromorphic functions corresponding to formal Laurent series," *SIAM J. Math. Anal.*, **10**, 1–17.

Jones, W. B. and Thron, W. J. (1980), *Continued Fractions, Analytic Theory and Applications*, Addison-Wesley, Reading, Mass.

Jones, W. B., Thron, W. J., and Waadeland, H. (1980), "A strong Stieltjes moment problem," *Trans. Am. Math. Soc.*, **261**, 503–528.

Joyce, D. C. (1971), "Survey of extrapolation processes in numerical analysis," *SIAM Rev.*, **13**, 435–490.

Joyce, G. S. and Guttmann, A. J. (1973), "A new method of series analysis," in P. R. Graves-Morris (ed.), *Padé Approximants and Their Applications*, Academic Press, London, pp. 163–167.

Kahaner, D. K. (1972), "Numerical quadrature by the ε-algorithm," *Math. Comp.*, **26**, 689–694.

Kailath, T., Vieira, A., and Morf, M. (1978), "Inverses of Toeplitz operators, innovations and orthogonal polynomials," *SIAM Review*, **20**, 106–119.

Karlin, S. (1968), *Total Positivity I*, Stanford Univ. Press.

Karlsson, J. (1976), "Rational interpolation and best rational interpolation," *J. Math. Anal. Appl.*, **53**, 38–52.

Karlsson, J. and von Sydow, B. (1976), "The convergence of Padé approximants to series of Stieltjes," *Arkiv for Matematik*, **14**, 43–53.

Karlsson, J. and Wallin, H. (1977), "Rational approximation by an interpolation procedure in several variables," in E. B. Saff and R. H. Varga (eds.), *Padé and Rational Approximation*, Academic Press, New York, pp. 83–100.

Khalil, A. B. (1977), "The K-matrix in distorted wave theory," *Nuovo Cim.*, **36A**, 354–366.

Killingbeck, J. (1977), "Quantum mechanical perturbation theory," *Rep. Prog. Phys.*, **40**, 963–1031.

Killingbeck, J. (1980), "The harmonic oscillator with λx^m perturbation," *J. Phys. A*, **13**, 49–56.

Kogbetlianz, E. G. (1960), "Generation of elementary functions," in A. Ralston and H. S. Wilf (eds.), *Mathematical Methods for Digital Computers*, Wiley, pp. 7–35.

Kronecker, L. (1881), "Zur Theorie der Elimination einer Variabeln aus zwei algebraischen Gleichungen," *Monatsberichte der Königlich Preussichen Akademie der Wissenschaften zu Berlin*, 535–600.

Laguerre, E. (1879), "Sur la réduction en fractions continues d'une fonction qui satisfait à une équation linéaire du premier ordre à coefficients rationnels," *Bull. Soc. Math de France*, **8**, 21–27.

Laguerre, E. (1885), "Sur la réduction en fractions continues d'une fraction qui satisfait à une équation différentielle linéaire du premier ordre dont les coefficients sont rationels," *J. de Math, Pures et Appliqués*, **1**, 135–165.

Lambert, J. D. (1974), "Two unconventional classes of methods for stiff systems," in R. A. Willoughby (ed.), *Stiff Differential Equations*, pp. 171–186.

Lambert, J. D. (1975), *Computational Methods in Ordinary Differential Equations*, Wiley.

Lambert, J. D. and Shaw, B. (1965), "On the numerical solution of $y'=f(x, y)$ by a class of formulae based on rational approximation," *Math. Comp.*, **19**, 456–462.

Lambert, J. D. and Shaw, B. (1966), "A generalization of multistep methods for ordinary differential equations," *Num. Math.*, **8**, 250–263.

Lanczos, C. (1952), "Solution of systems of linear equations by minimized iterations," *J. Res. Nat. Bur. Stand.*, **49**, 33–53.

Lanczos, C. (1957), *Applied Analysis*, Pitman Press.

Larkin, F. M. (1967), "Some techniques for rational interpolation," *Comp. J.*, **10**, 178–187.

Larkin, F. M. (1981), "Rootfinding by divided differences," *Num. Math.*, **37**, 93–104.

Lee, T. D. and Yang, C. N. (1952), "Statistical theory of equations of state and phase transitions II. Lattice gas and Ising model," *Phys. Rev.*, **87**, 410–419.

Leighton, W. and Scott, W. T. (1939), "A general continued fraction expansion," *Bull. Am. Math. Soc.*, **45**, 596–605.

Levin, D. (1973), "Development of non-linear transformations for improving convergence of sequences," *Int. J. Comp. Math.*, **B3**, 371–388.

Levin, D. (1976), "General order Padé type rational approximants defined from double power series," *J. Inst. Math. Appl.*, **18**, 1–8.

Loeffel, J. J., Martin, A., Simon, B., and Wightman, A. S. (1969), *Phys. Lett.*, **30B**, 656–658.

Longman, I. M. (1966), "The application of rational approximations to the solution of problems in theoretical seismology," *Bull. Seismol. Soc. America*, **56**, 1045–1065.

Longman, I. M. (1972), "Numerical Laplace transform inversion of a function arising in viscoelasticity," *J. Comp. Phys.*, **10**, 224–231.

Longman, I. M. (1973), "Use of Padé table for approximate Laplace transform inversion," in P. R. Graves-Morris (ed.), *Padé Approximants and Their Applications*, Academic Press, London, pp. 131–134.

Longman, I. M. (1974), "Best rational function approximation for Laplace transform inversion," *SIAM J. Math. Anal.*, **5**, 574–580.

Lopez, C. and Yndurain, F. J. (1973), "Model independent calculation of Kp dispersion relations; extrapolation with Padé techniques," *Nucl. Phys. B*, **64**, 315–333.

Lovelace, C. (1964), "Practical theory of three-particle states I. Non-relativistic," *Phys. Rev.*, **135**, B1225–1249.

Lovelace, C. and Masson, D. (1962), "Calculation of Regge poles by continued fractions," *Nuovo Cim.*, **26**, 472–484.

Lubinsky, D. S. (1980), "Exceptional Sets of Padé approximants," Ph.D. thesis, Univ. of Witwatersrand.

Lubkin, S. (1952), "A method of summing infinite series," *J. Res. Nat. Bur. Stand.*, **48**, 228–254.

Luke, Y. L. (1958), "The Padé table and the τ-method," *J. Math and Phys.*, **37**, 110–127.

Luke, Y. L. (1960), "On the economic representations of transcendental functions," *J. Math. and Phys.*, **38**, 279–294.

Luke, Y. L. (1962), "On the approximate inversion of some Laplace transforms," in *Proc. 4th U.S. Nat. Congr. Appl. Mech.*, pp. 269–276.

Luke, Y. L. (1970), "Evaluation of the gamma function by means of Padé approximations," *SIAM J. Num. Anal.*, **1**, 266–281.

Luke, Y. L. (1975), "On the error in the Padé approximants for a form of incomplete gamma function including the exponential," *SIAM J. Math. Anal.*, **6**, 829–839.

Luke, Y. L. (1977), "On the error in Padé approximations for functions defined by Stieltjes integrals," *Comp. and Math. with Appl.*, **3**, 307–314.

Luke, Y. L. (1978), "Error estimation in numerical inversion of Laplace transform using Padé approximation," *J. Franklin Inst.*, **305**, 259–273.

Luke, Y. L. (1980), "Computations of coefficients in the polynomials of Padé approximations by solving systems of linear equations," *J. Comp. and Appl. Math.*, **6**, 213–218.

Luke, Y. L., Fair, W., and Wimp, J. (1975), "Predictor-corrector formulas based on rational interpolants," *Int. J. Computers and Math. with Appl.*, **1**, 3–12.

Lutterodt, C. L. (1974), "A two-dimensional analogue of Padé approximant theory," *J. Phys. A*, **7**, 1027–1037.

Lyness, J. and Ninham, B. W. (1967), "Numerical quadrature and asymptotic expansions," *Math. Comp.*, **21**, 162–178.

McCabe, J. H. (1974), "A continued fraction expansion, with a truncation estimate, for Dawson's integral," *Math. Comp.*, **28**, 811–816.

McCabe, J. H. (1975), "A formal extension of the Padé table to include two point Padé quotients," *J. Inst. Math. Appl.*, **15**, 363–372.

McCabe, J. H. and Murphy, J. A. (1976), "Continued fractions which correspond to power series at two points," *J. Inst. Math. Appl.*, **17**, 233–247.

McCleod, J. B. (1971), "A note on the ε-algorithm," *Computing*, **7**, 17–24.

McCoy, B. M. and Wu, T. T. (1973), *The Two Dimensional Ising Model*, Harvard Univ. Press, Cambridge, Mass.

McEliece, R. J. and Shearer, J. B. (1978), "A property of Euclid's algorithm and its application to Padé approximation," *SIAM J. Appl. Math.*, **34**, 611–615.

Maehly, H. J. (1956), October monthly progress report, Institute for Advanced Study, Princeton.

Maehly, H. J. (1960), "Rational approximations for transcendental functions," in *Proc. Int. Conf. on Inform. Proc., 1959*, Butterworth, pp. 57–62.

Maehly, H. J. and Witzgall, Ch. (1960), "Tschebyscheff–Approximationen in kleinen Intervallen II. Stetigkeitssätze für gebrochen rationale Approximationen," *Num. Math.*, **2**, 293–307.

Magnen, J. and Sénéor, R. (1976), "The infinite volume limit of the ϕ_3^4 model," *Ann. d'Inst. H. Poincaré A*, **24**, 95–159.

Magnus, A. (1962), "Expansion of power series into P-functions," *Math. Z.*, **80**, 209–216.

Markov, A. (1884), "Démonstration de certaines inégalités de Tchebychef," *Math. Ann.*, **24**, 172–180.

Marx, I. (1963), "Remark concerning a non-linear sequence to sequence transformation," *J. Math. and Phys.*, **42**, 334–335.

Mason, J. C. (1967), "Chebychev polynomial approximations for the L-membrane eigenvalue problem," *SIAM J. Appl. Math.*, **15**, 172–181.

Masson, D. (1967a), "Padé approximant and the partial-wave integral equation," *J. Math. Phys.*, **8**, 512–514.

Masson. D. (1967b), "Analyticity in the potential strength," *J. Math. Phys.*, **8**, 2308–2314.

Meinguet, J. (1970), "On the solubility of the Cauchy interpolation problem," in A. Talbot (ed.), *Approximation Theory*, Academic Press, London, pp. 137–163.

Mery, P. (1977), "A variational approach to operator and matrix Padé approximation. Applications to potential scattering and field theory," in E. B. Saff and R. H. Varga (eds.), *Padé and Rational Approximation*, Academic Press, New York, pp. 375–388.

Merz, G. (1968), "Padésche Näherungsbrüche und Iterationsverfahren höherer Ordnung," *Comp.*, **3**, 165–183.

Michalik, B. (1970), "Distortion operator method and Padé approximants," *Ann. Phys.*, **57**, 201–213.

Mills, W. H. (1975), "Continued fraction and linear recurrences," *Math. Comp.*, **29**, 173–180.

Milne-Thomson, L. M. (1960), *Calculus of Finite Differences*, Macmillan, London.

de Montessus de Ballore, R. (1902), "Sur les fractions continues algébriques," *Bull. Soc. Math. de France*, **30**, 28–36.

de Montessus de Ballore, R. (1905), "Sur les fractions continues algébriques," *Rend. di Palermo*, **19**, 1–73.

Muir, T. (1960), *A Treatise on the Theory of Determinants*, Dover, New York.

Murphy, J. A. (1971), "Certain rational fraction approximations to $(1+x^2)^{-1/2}$," *J. Inst. Math. Appl.*, **7**, 138–150.

Murphy, J. A. and O'Donohoe, M. R. (1977), "A class of algorithms for rational approximation of functions formally defined by power series," *J. Appl. Math. and Phys.* (*ZAMP*), **28**, 1121–1131.

Narcowich, F. J. and Allen, G. D. (1975), "Convergence of diagonal operator-valued Padé approximants to Dyson expansion," *Comm. Math. Phys.*, **45**, 153–157.

Nelson, E. (1973), "Construction of quantum fields from Markoff fields," *J. Funct. Anal.*, **12**, 97–112.

Newton, R. G. (1966), *Scattering Theory of Waves and Particles*, McGraw-Hill.

Ninham, B. W. and Lyness, J. (1969), "Asymptotic expansions for the error functional," *Math. Comp.*, **23**, 71–83.

Nørsett, S. P. (1974), "One step methods of Hermite type for numerical integration of stiff systems," *B.I.T.*, **14**, 63–77.

Nuttall, J. (1966), "Padé approximants and bounds on the Bethe–Salpeter amplitude," *Phys. Lett.*, **23**, 492.

Nuttall, J. (1967), "Convergence of Padé approximants for the Bethe–Salpeter amplitude," *Phys. Rev.*, **157**, 1312–1316.

Nuttall, J. (1970a), "Connection of Padé approximants with stationary variational principles and the convergence of certain Padé approximants," in G. A. Baker, Jr. and J. L. Gammel (eds.), *The Padé Approximant in Theoretical Physics*, Academic Press, New York, pp. 219–230.

Nuttall, J. (1970b), "Convergence of Padé approximants of meromorphic functions," *J. Math. Anal. Appl.*, **31**, 147–153.

Nuttall, J. (1973), "Variational principles and Padé approximants," in P. R. Graves-Morris (ed.), *Padé Approximants and Their Applications*, Academic Press, New York, pp. 29–40.

Nuttall, J. (1977), "The convergence of Padé approximants to functions with branch points," in E. B. Saff and R. H. Varga (eds.), *Padé and Rational Approximation*, Academic Press, New York, pp. 101–109.

Nuttall, J. (1980), "Sets of minimum capacity, Padé approximants and the bubble problem," in C. Bardos and D. Bessis (eds.), *Bifurcation Problems in Mathematical Physics and Related Topics*, D. Reidel, Boston, pp. 185–201.

Nuttall, J. and Singh, S. R. (1977), "Orthogonal polynomials and Padé approximants associated with a system of arcs," *J. Approx. Theory*, **21**, 1–42.

Ostrowski, A. (1925), "Uber Folgen analytischer Funktionen und einige Verschärfungen des Picarden Satzes," *Math. Zeit.*, **24**, 215–258.

Padé, H. (1892), "Sur la représentation approchée d'une fonction par des fractions rationelles," *Ann. de l'Ecole Normale Sup. $3^{iéme}$ Série*, **9**, Suppl., 3–93.

Padé, H. (1894), "Sur les séries entières convergents on divergents, et les fractions continues rationelles," *Acta Math.*, **18**, 97–112.

Padé, H. (1899), "Memoire sur les développements en fractions continues de la fonction exponentielle pouvant servir d'introduction à la théorie des fractions continues algébriques," *Ann. Sci. Ecole Norm. Sup.*, **16**, 395–426.

Padé, H. (1900), "Sur la distribution des réduites anormales d'une fonction," *Comptes Rendus*, **130**, 102–104.

Padé, H. (1901), "Sur l'expression générale de la fonction rationelle approchée de $(1 + x)^m$," *Comptes Rendus*, **132**, 754–756.

Padé, H. (1907), "Recherches sur la convergence de développements en fractions continues d'une certaine catégorie de fonctions," *Ann. Sci. Ecole Norm. Sup.*, **24**.

Paydon, J. F. and Wall, H. S. (1942), "The continued fraction as a sequence of linear transformations," *Duke Math. J.*, **9**, 360–372.

Pekeris, C. L. (1958), "Ground state of two electron atoms," *Phys. Rev.*, **112**, 1649–1658.

Pekeris, C. L. (1959), "Binding energy of helium," *Phys. Rev.*, **115**, 1216–1221.

Peres, A. (1963), "Mechanical model for quantum field theory," *J. Math. Phys.*, **4**, 332–333.

Perron, O. (1957), *Die Lehre von den Kettenbrüchen*, B. G. Teubner, Stuttgart.

Pfeuty, P., Jasnow, D., and Fisher, M. E. (1974), "Crossover scaling functions for exchange anisotropy," *Phys. Rev. B*, **10**, 2088–2112.

Pindor, M. (1976), "A simplified algorithm for calculating the Padé table derived from the Baker and Longman schemes," *J. Comp. Appl. Math.*, **2**, 255–258.

Pindor, M. (1979a), "Padé approximants and rational functions as tools for finding poles and zeros of analytical functions measured experimentally," in L. Wuytack (ed.), *Padé Approximation and Its Applications*, Academic Press, New York, pp. 338–347.

Pindor, M. (1979b), "Matrix variational Padé approximants and multistep square well potentials," *Lett. Math. Phys.*, **3**, 223–228.

Pindor, M. (1980), "Unambiguous results from variational matrix Padé approximants," CERN report.

Pommerenke, Ch. (1973), "Padé approximants and convergence in capacity," *J. Math. Anal. Appl.*, **41**, 775–780.

Pringsheim, A. (1910), "Uber Konvergenz und Funktionen-theoretischen Charakter gewisser limitärperiodischer Kettenbrüche," *Sitzungsberichte der Math.-Physikalische Klasse der Kgl. Bayerischen Akademie der Wissenschaften zu München*, **6**, 1–52.

Reed, M. and Simon, B. (1979), "Scattering Theory," vol. 3 in *Methods of Modern Mathematical Physics*, Academic Press, New York.

Rehr, J. J., Joyce, G. S., and Guttmann, A. J. (1980), "A recurrence technique for confluent singularity analysis of power series," *J. Phys. A*, **13**, 1587–1602.

Reid, W. T. (1972), *Riccati Equations*, Academic Press, London.

Richardson, L. F. (1927), "The deferred approach to the limit, I. Single lattice," *Phil. Trans. Roy. Soc. Lond.*, **A226**, 299–349.

Riesz, F. and Nagy, B. v. Sz. (1955), *Functional Analysis*, Ungar.

Riesz, M. (1922a), "Sur le problème des moments," *Arkiv för Matematik, Astronomi och Fysik*, **16** (12), 1–23.

Riesz, M. (1922b), "Sur le problème des moments," *Arkiv för Matematik, Astronomi och Fysik*, **16** (19), 1–21.

Riesz, M. (1923), "Sur le problème des moments," *Arkiv för Matematik, Astronomi och Fysik*, **17**, (16), 1–52.

Riordan, J. (1966), *Combinatorial Identities*, Wiley.

Rissanen, J. (1973), "Algorithms for triangular decomposition of block Hankel and Toeplitz matrices with applications to factoring positive matrix polynomials," *Math. Comp.*, **27**, 147–155.

Rivlin, T. J. (1969), *An Introduction to the Theory of Functions*, Blaisdell.

Roberts, D. E. (1977), "An analysis of double power series using rotationally covariant approximants," *J. Comp. Appl. Math.*, **3**, 257–262.

Roberts, D. E., Griffiths, H. P., and Wood, D. W. (1975), "The analysis of double power series using Canterbury approximants," *J. Phys. A*, **8**, 1365–1372.

Robinson, P. D. and Barnsley, M. F. (1979), "Pointwise bivariational bounds on solutions of Fredholm integral equations," *SIAM J. Num. Anal.*, **16**, 135–144.

Rogers, L. J. (1907), "On the representation of certain asymptotic series as convergent continued fractions," *Proc. Lond. Math. Soc.*, **4**, 72–89.

Roth, A. (1938), "Approximationseigenschaften und Strahlengrenzwerte unendlich vieler linearer Gleichungen," *Comm. Math. Helv.*, **11**, 77–125.

Roth, R. (1965), "The qualifying examination," *Math. Mag.*, **38**, 166–167.

Rudin, W. (1976), *"Principles of Mathematical Analysis,"* McGraw-Hill.

Ruijgrok, Th. W. (1968), "A new formulation of the theory of strong coupling," *Nucl. Phys. B*, **8**, 591–608.

Ruijgrok, Th. W. (1972), "A numerical study of analyticity in the coupling constant," *Nucl. Phys. B*, **39**, 616–642.

Runge, C. (1885) "Zur Theorie der eindeutigen analytischen Funktionen" *Acta Math.*, **6**, 229–244.

Rutishauser, H. (1954), "Der Quotienten-Differenzen-Algorithmus," *Zeit. für Angewandte Math.*, **5**, 233–251.

Rutishauser, H. (1957), *Der Quotienten-Differenzen-Algorithmus*, Birkhäuser Verlag, Basel.

Saff, E. B. (1972), "An extension of Montessus de Ballore's theorem on the convergence of interpolating rational functions," *J. Approx. Theory*, **6**, 63–67.

Saff, E. B., Schönhage, A., and Varga, R. S. (1978), "Geometrical convergence to e^{-z} by rational functions with real poles," *Num. Math.*, **25**, 307–322.

Saff, E. B. and Varga, R. S. (1975), "On the zeros and poles of Padé approximants to e^z," *Num. Math.*, **25**, 1–14.

Saff, E. B. and Varga, R. S. (1976a), "Zero-free parabolic regions for sequences of polynomials," *SIAM J. Math. Anal.*, **7**, 344–357.

Saff, E. B. and Varga, R. S. (1976b), "On the sharpness of theorems concerning zero free regions for certain sequences of polynomials," *Num. Math.*, **26**, 345–354.

Saff, E. B. and Varga, R. S. (1977), "On the zeros and poles of Padé approximants to e^z. II," in E. B. Saff and R. S. Varga (eds.), *Padé and Rational Approximation*, Academic Press, New York, pp. 195–214.

Saff, E. B. and Varga, R. S. (1978), "Non-uniqueness of best rational approximations to real functions on real intervals," *J. Approx. Theory*, **23**, 78–85.

Saff, E. B., Varga, R. S., and Ni, W.-C. (1976), "Geometric convergence of rational approximations to e^{-z} in infinite sectors," *Num. Math.*, **26**, 211–225.

Salzer, H. E. (1962), "Note on osculatory rational interpolation," *Math. Comp.*, **16**, 486–491.

Schlessinger, L. and Schwartz, C. (1966), "Analyticity as a computational tool," *Phys. Rev. Lett.*, **16**, 1173–1174.

Schoenberg, I. J. (1951), "On the Pólya frequency functions I: The totally positive functions and their Laplace transforms," *J. d'Anal. Math.*, **1**, 331–374.

Schofield, D. F. (1972), "Continued fraction method for perturbation theory," *Phys. Rev. Lett.*, **29**, 811–814.

Schwartz, C. (1966), "Information content of the Born series," *J. Comp. Phys.*, **1**, 21–28.

Schwartz, C. and Zemach, C. (1966), "Theory and calculation with the Bethe–Salpeter equation," *Phys. Rev.*, **141**, 1454–1467.

Schweber, S. S. (1961), *Introduction to Relativistic Quantum Field Theory*, Harper and Row, New York.

Scott, W. T. and Wall, H. S. (1940a), "Continued fraction expansions for arbitrary power series," *Ann. Math.*, **41**, 328–349.

Scott, W. T. and Wall, H. S. (1940b), "A convergence theorem for continued fractions," *Trans. Am. Math. Soc.*, **47**, 155–172.

Shafer, R. E. (1974), "On quadratic approximation," *SIAM J. Num. Anal.*, **11**, 447–460.

Shanks, D. (1955), "Non-linear transformations of divergent and slowly convergent sequences," *J. Math. Phys.*, **34**, 1–42.

Sheludyak, Yu. E. and Rabinovich, V. A. (1979), "On the values of the critical indices of the three dimensional Ising model," *High Temp.*, **17**, 40–43.

Short, L. (1978), "The practical evaluation of multivariate approximants with branch points," *Proc. Roy. Soc. London A*, **362**, 57–69.

Sidi, A. (1979), "Convergence properties of some non-linear sequence transformations," *Math. Comp.*, **33**, 315–326.

Sidi, A. (1980a), "Some aspects of two-point Padé approximants," *J. Comp. Appld. Math.*, **6**, 9-17.

Sidi, A. (1980b), "Analysis of the convergence of the T-transformation for power series," *Math. Comp.*, **35**, 833–850.

Siemieniuch, J. L. (1976), "Properties of certain rational approximations to e^{-z}," *B.I.T.*, **16**, 172–191.

Simon, B. (1970), "Coupling constant analyticity for the anharmonic oscillator," *Ann. Phys.*, **58**, 76–136.

Simon, B. (1972), "The anharmonic oscillator—a singular perturbation theory," in D. Bessis (ed.), *Cargèse Lecture Notes in Physics*, Vol. 5, Gordon and Breach, pp. 383–414.

Smith, D. A. and Ford, W. F. (1979), "Acceleration of linear and logarithmic convergence," *SIAM J. Num. Anal.*, **16**, 223–240.

Smith, I. M., Siemieniuch, J. L., and Gladwell, I. (1977), "Evaluating Nørsettmethods for integrating differential equations in time,"*Int. J. Num Anal. Meth. in Geomechanics*, **1**, 57–74.

Smithies, F. (1958), *Integral Equations*, Cambridge University Press.

Starkand, Y. (1976), "Subroutine for calculation of matrix Padé approximants," *Comm. Comp. Phys.*, **11**, 325–330.

Stern, M. S. and Warburton, A. E. A. (1972), "Finite difference solution of the partial wave Schrödinger equation," *J. Phys. A*, **5**, 112–124.

Stieltjes, T. J. (1884), "Quelques recherches sur les quadratures dites méchaniques," *Ann. Sci. Ecole. Norm. Sup.*, **1**, 409–426.

Stieltjes, T. J. (1889), "Sur la réduction en fraction continue d'une série précédant suivant les puissances descendents d'une variable," *Ann. Fac. Sci. Toulouse*, **3H**, 1–17.

Stieltjes, T. J. (1894), "Recherches sur les fractions continues," *Ann. Fac. Sci. Toulouse*, **8J**, 1–122; **9A**, 1–47.

Stoer, J. (1961), "Über zwei Algorithmen zur Interpolation mit rationalen Funktionen," *Num. Math.*, **3**, 285–304.

Stone, M. H. (1932), *Linear Transformations in Hilbert Space*, Am. Math. Soc., Providence.

Takehasi and Mori, (1971), "Estimation of errors in the numerical quadrature of analytic functions," *Applicable Analysis*, **1**, 201–229.

Talbot, A. (1979), "The accurate numerical inversion of Laplace transforms," *J. Inst. Math. Appl.*, **23**, 97–120.

Tani, S. (1965), "Padé approximants in potential scattering," *Phys. Rev.*, **139**, B1011–1020.

Tani, S. (1966a), "Complete continuity of the kernel in generalized potential scattering I. Short range interaction without strong singularity," *Ann. Phys.*, **37**, 411–450.

Tani, S. (1966b), "Complete continuity of the kernel in generalized potential scattering II. Generalized Fourier series expansion," *Ann. Phys.*, **37**, 451–486.

Taylor, J. M. (1978), "The condition of Gram matrices and related problems," *Proc. Roy. Soc. Edin.*, **80A**, 45–56.

Tchebycheff, P. (1858), "Sur les fractions continues," *J. de Math.*, **8**, 289–323.

Tchebycheff, P. (1874), "Sur les valeurs limites des intégrales," *J. de Math. Pures et Appl.*, **19**, 157–160.

Thacher, H. C. and Tukey, J. (1960), "Rational interpolation made easy by a recursive algorithm," unpublished manuscript.

Thacher, H. C. (1974), "Numerical application of the generalized Euler transformation," in J. L. Rosenfeld (ed.), *Information Processing 74*, North Holland, Amsterdam, pp. 627–631.

Thiele, T. N. (1909), *Interpolationsrechnung*, Teubner.

Thompson, C. J., Guttman, A. J., and Ninham, B. W. (1969), "Determination of critical behaviour in lattice statistics from series expansions, II," *J. Phys. C*, **2**, 1889–1899.

Thron, W. J. (1948), "Some properties of continued fractions $1 + d_0 z + K(z/(1 + d_n z))$," *Bull. Am. Math. Soc.*, **54**, 206–218.

Thron, W. J. (1961), "Convergence regions for continued fractions and other infinite processes," *Am. Math. Monthly*, **68**, 734–750.

Thron, W. J. (1974), "A survey of recent convergence results for continued fractions," *Rocky Mtn. J. Math.*, **4**, 273–282.

Thron, W. J. (1977), "Two point Padé tables, *T*-fractions and sequences of Schur," in E. B. Saff and R. S. Varga (eds.), *Padé and Rational Approximation*, Academic Press, New York, pp. 215–225.

Titchmarsh, E. C. (1939), *Theory of Functions*, Oxford University Press.

Tjon, J. A. (1970), "The Padé approximant in three body calculations," *Phys. Rev. D*, **1**, 2109–2112.

Tjon, J. A. (1973), "Application of Padé approximants in the three body problem," in P. R. Graves-Morris (ed.), *Padé Approximants*, Inst. of Phys., Bristol, pp. 241–252.

Tjon, J. A. (1977), "Operator Padé approximants and three body scattering," in E. B. Saff and R. S. Varga (eds.), *Padé and Rational Approximation*, Academic Press, New York, pp. 389–396.

Traub, J. F. (1964), *Iterative Methods for the Solution of Equations*, Prentice-Hall, Englewood Cliffs, N.J.

Trench, W. (1964), "An algorithm for the inversion of finite Toeplitz matrices," *SIAM J. Appl. Math.*, **12**, 515–522.

Trench, W. (1965), "An algorithm for the inversion of finite Hankel matrices," *SIAM J. Appl. Math.*, **13**, 1102–1107.

Tricomi, F. G. (1957), *Integral Equations*, Interscience.

Trudi, N. (1862), *Teoria de' Determinati e Loro Applicazioni*, Libreria Scientifica e Industriale de B. Pellerano, Napoli.

Turchetti, G. (1976), "Variational principles and matrix approximants in potential theory," *Lett. Nuovo Cim.*, **15**, 129–133.

Turchetti, G. (1978), "Variational matrix Padé approximants in two body scattering," *Fortsch. Physik*, **26**, 1–28.

Van Rossum, H. (1971), "On the poles of Padé approximants to e^z," *Nieuw Archief voor Wiskunde*, **29**, 37–45.

Van Vleck, E. B. (1903), "On an extension of the 1894 memoir of Stieltjes," *Trans. Am. Math. Soc.*, **4**, 297–332.

Van Vleck, E. B. (1904), "On the convergence of algebraic continued fractions whose coefficients have limiting values," *Trans. Am. Math. Soc.*, **5**, 253–262.

Varga, R. S. (1961), "On higher order stable implicit methods for solving parabolic partial differential equations," *J. Math. Phys.*, **40**, 220–231.

Varga, R. S. (1962), *Matrix Iterative Analysis*, Prentice-Hall, Englewood Cliffs, N.J.

Villani, M. (1972), "A summation method for perturbative series with divergent terms," in D. Bessis (ed.), *Cargèse Lecture Notes in Physics*, Vol. 5, Gordon and Breach, New York, pp. 461–474.

Viskovatov, B. (1803–1806), "De la méthode générale pour réduire toutes sortes des quantités en fractions continues," *Memoires de L'Academie Impériale des Sciences de St. Petersburg*, **1**, 226–247.

Vorobyev, Yu. V. (1965), *Method of Moments in Applied Mathematics*, Gordon and Breach, New York.

Waadeland, H. (1979), "General *T*-fractions corresponding to functions satisfying certain boundedness conditions," *J. Approx. Theory*, **26**, 317–328.

Wall, H. S. (1931), "On the Padé approximants associated with a positive definite power series," *Trans. Am. Math. Soc.*, **33**, 511–532.

Wall, H. S. (1932a), "On the relationship among the diagonal files of a Padé table," *Bull. Am. Math. Soc.*, **38**, 752–760.

Wall, H. S. (1932b), "General theorems on the convergence of sequences of Padé approximants," *Trans. Am. Math. Soc.*, **34**, 409–416.

Wall, H. S. (1945), "Note on a certain continued fraction," *Bull. Am. Math. Soc.*, **51**, 930–934.

Wall, H. S. (1948), *The Analytic Theory of Continued Fractions*, Van Nostrand, Princeton, N.J.

Wallin, H. (1972), "On the convergence theory of Padé approximants," in P. L. Butzer, J.-P. Kahane, and B. Sz-Nagy (eds.), *Linear Operators and Approximation* (Proc. Conf. at Oberwolfach), Birkhäuser, Basel, pp. 461–469.

Wallin, H. (1974), "The convergence of Padé approximants and the size of the power series coefficients," *Appl. Anal.*, **4**, 235–251.

Walsh, J. L. (1964a), "Convergence of sequences of rational functions of best approximation. I," *Math. Ann.*, **155**, 252–264.

Walsh, J. L. (1964b), "Padé approximants as limits of rational functions of best approximation," *J. Math. and Mech.*, **13**, 305–312.

Walsh, J. L. (1965a), "Convergence of sequences of rational functions of best approximation II," *Trans. Am. Math. Soc.*, **116**, 227–237.

Walsh, J. L. (1965b), "The convergence of sequences of rational functions of best approximation with some free poles," in H. L. Garabedian (ed.), *Approximation of Functions*,
Elsevier, Amsterdam, pp. 1–16.

Walsh, J. L. (1967), "On the convergence of sequences of rational functions," *SIAM J. Num.
Anal.*, **4**, 211–221.

Walşh, J. L. (1969), "Interpolation and approximation by rational functions in the complex
domain," *Am. Math. Soc. Colloq. Pub.*, Vol. 20, Providence, R.I.

Walsh, J. L. (1970), "Approximation by rational functions: open problems," *J. Approx. Theory*,
3, 236–242.

Warburton, A. E. A. and Stern, M. S. (1968), "Two body off-shell potential scattering," *Nuovo
Cim.*, **60A**, 131–159.

Warner, D. D. (1976), "An extension of Saff's theorem on the convergence of interpolating
rational functions," *J. Approx. Theory*, **18**, 108–118.

Watson, P. J. S. (1973), "Algorithms for differentiation and integration," in P. R. Graves-Morris
(ed.), *Padé Approximants and Their Applications*, Academic Press, London, pp. 93–98.

Watson, P. J. S. (1974), "Two variable rational approximants—a new method," *J. Phys. A*, **7**,
*L*167–*L*170.

Werner, H. (1979), "A reliable method for rational interpolation," in L. Wuytack (ed.), *Padé
Approximation and Its Applications*, Springer Lecture Notes in Mathematics, No. 765,
pp. 257–277.

Werner, H. (1980), "Ein algorithmus zur rationalen Interpolation," in *Numerische Methoden der
Approximationstheorie* **5**, eds. L. Collatz, G. Meinardus and H. Werner, Birkhäuser,
319–337.

Werner, H. and Schaback, R. (1972), *Praktische Mathematik II*, Springer-Verlag, Berlin.

Werner, H. and Wuytack, L. (1978), "Nonlinear quadrature rules in the presence of a
singularity," *Comp. & Math. with Applcns.*, **4**, 237–245.

Wetterling, W. (1963), "Ein Interpolationsverfahren zur Lösung der linearen Gleichungssys-
teme, die bei der rationalen Tchebyscheff-Approximation auftreten," *Archiv. Rat. Mech.
und Anal.*, **12**, 403–408.

Wheeler, J. C. and Gordon, R. G. (1970), "Bounds for averages using moment constraints," in
G. A. Baker, Jr., and J. L. Gammel (eds.), *The Padé Approximant in Theoretical Physics*,
Academic Press, New York, pp. 99–128.

Widder, D. V. (1972), *The Laplace Transform*, Princeton University Press.

Wilkinson, J. H. (1963), *Rounding Error in Algebraic Processes*, Notes in Applied Science, No.
32, H.M.S.O., London.

Wilkinson, J. H. (1965), *The Algebraic Eigenvalue Problem*, Oxford University Press.

Wilson, R. (1927), "Divergent continued fractions and polar singularities," *Proc. Lond. Math.
Soc.*, **26**, 159–168.

Wilson, R. (1928a), "Divergent continued fractions and polar singularities II. Boundary pole
multiple," *Proc. Lond. Math. Soc.*, **27**, 497–512.

Wilson, R. (1928b), "Divergent continued fractions and polar singularities III. Several boundary
poles," *Proc. Lond. Math. Soc.*, **28**, 128–144.

Wilson, R. (1930), "Divergent continued fractions and non-polar singularities," *Proc. Lond.
Math. Soc.*, **30**, 38–57.

Wood, D. W. and Griffiths, H. P. (1974), "Chisholm approximants and critical phenomena," *J.
Phys. A*, **7**, *L*101–*L*104.

Wright, K. (1970), "Some relationships between implicit Runge-Kutta, collocation and Lanczos τ-methods and their stability properties," *B.I.T.*, **10**, 217–227.

Wuytack, L. (1973), "An algorithm for rational interpolation similar to the q.d.-algorithm," *Num. Math.*, **20**, 418–424.

Wuytack, L. (1974a), "Numerical integration by using non-linear techniques," *J. Comp. Appl. Math.*, **1**, 261–272.

Wuytack, L. (1974b), "On some aspects of the rational interpolation problem," *SIAM J. Num. Anal.*, **11**, 52–60.

Wuytack, L. (1975), "On the osculatory rational interpolation problem," *Math. Comp.*, **29**, 837–843.

Wuytack, L. (1979), "Commented bibliography," in L. Wuytack (ed.), *Padé Approximation and Its Applications*, Springer Lecture Notes in Mathematics, No. 765, pp. 375–392.

Wynn, P. (1956), "On a device for calculating the $e_m(S_n)$ transformation," *Math. Tables and A.C.*, **10**, 91–96.

Wynn, P. (1960), "The rational approximation of functions which are formally defined by power series expansions," *Math. Comp.*, **14**, 147–186.

Wynn, P. (1961a), "The epsilon algorithm and operational formulas of numerical analysis," *Math. Comp.*, **15**, 151–158.

Wynn, P. (1961b), "L'ε-algoritmo e la tavola di Padé," *Rendi. Mat. Roma*, **20**, 403–408.

Wynn, P. (1962a), "The numerical efficiency of certain continued fraction expansions," *Proc. Kon. Ned. Akad. Wetensch. A*, **65**, 127–154.

Wynn, P. (1962b), "Acceleration techniques for iterated vector and matrix problems," *Math. Comp.*, **16**, 301–322.

Wynn, P. (1963), "Continued fractions whose coefficients obey a non-commutative law of multiplication," *Arch. Rat. Mech. Anal.*, **12**, 273–312.

Wynn, P. (1964), "General purpose vector epsilon algorithm procedures," *Num. Math.*, **6**, 22–36.

Wynn, P. (1966a), "Upon systems of recursions which obtain among quotients of the Padé table," *Num. Math.*, **8**, 264–269.

Wynn, P. (1966b), "On the convergence and stability of the ε-algorithm," *SIAM J. Num. Anal.*, **3**, 91–122.

Wynn, P. (1967), "A general system of orthogonal polynomials," *Quart. J. Math.*, **18**, 81–96.

Wynn, P. (1968), "Upon the Padé table derived from a Stieltjes series," *SIAM J. Num. Anal.*, **5**, 805–834.

Wynn, P. (1973), "On the zeros of certain confluent hypergeometric functions," *Proc. Am. Math. Soc.*, **40**, 173–182.

Wynn, P. (1974), "Some recent developments in the theories of continued fractions and the Padé table," *Rocky Mtn. J. Math.*, **4**, 297–323.

Wynn, P. (1977), "The transformation of series by the use of Padé quotients and more general approximants," in E. B. Saff and R. S. Varga (eds.), *Padé and Rational Approximation*, Academic Press, New York, pp. 121–146.

Young, R. C., Biedenharn, L. C., and Feenberg, E. (1957), "Continued fraction approximants to the Brillouin-Wigner perturbation series," *Phys. Rev.*, **106**, 1151–1155.

Zinn-Justin, J. (1970), "Strong interaction dynamics with Padé approximants," *Phys. Repts.*, **1C**, 56–102.

Zinn-Justin, J. (1971), "Convergence of Padé approximants in the general case," in A. Visconti (ed.), *Colloquium on Advanced Computing Methods in Theoretical Physics*, CNRS, Marseille.

Zinn-Justin, J. (1973), "Recent developments in the theory of Padé approximants," in A. Visconti (ed.), *International Colloquium on Advanced Computing Methods in Theoretical Physics*, Vol. 2, CNRS, Marseille, p. C-XIII-1.

Zohar, S. (1974), "The solution of a Toeplitz set of linear equations," *J. Assoc. Comp. Mech.*, **21**, 272–276.

Selected Bibliography on Padé Approximants and Critical Phenomena

Baker, G. A., Jr. (1961), "Application of the Padé approximant method to the investigation of some magnetic properties of the Ising model," *Phys. Rev.*, **124**, 768–774.

Baker, G. A., Jr. (1967), "Convergent, bounding approximation procedures to the ferromagnetic Ising model," *Phys. Rev.*, **161**, 434–445.

Baker, G. A., Jr. (1973), "Generalized Padé approximant bounds for critical phenomena," in P. R. Graves-Morris (ed.), *Padé Approximants and Their Applications*, Academic Press, London, pp. 147–158.

Baker, G. A., Jr. (1975), "Self-interacting boson quantum field theory and the thermodynamic limit in d dimensions," *J. Math. Phys.*, **16**, 1324–1346.

Baker, G. A., Jr. (1977a), "Application of Padé approximants to critical phenomena," in E. B. Saff and R. S. Varga (eds.), *Padé and Rational Approximation*, Academic Press, New York, pp. 323–337.

Baker, G. A., Jr. (1977b), "Analysis of hyperscaling in the Ising model by the high temperature series method," *Phys. Rev. B*, **15**, 1552–1559.

Baker, G. A., Jr. (1979), "The continuous-spin Ising model of field theory and the renormalization group," in C. Bardos and D. Bessis (eds.), *Bifurcation Phenomena in Mathematical Physics and Related Topics*, D. Reidel, Boston.

Baker, G. A., Jr., Gilbert, H. E., Eve, J., and Rushbrooke, G. S. (1967), "High temperature expansion for the spin-$\frac{1}{2}$ Heisenberg model," *Phys. Rev.*, **164**, 800–817.

Baker, G. A., Jr., Nickel, B. G., Green, M. S., and Meiron, D. I. (1976), "Ising model critical indices in three dimensions from the Callan-Symanzik equation," *Phys. Rev. Lett.*, **36**, 1351–1354.

Baker, G. A., Jr., Nickel, B. G., and Meiron, D. I. (1978), "Critical indices from perturbation analysis of the Callan-Symanzik equation," *Phys. Rev. B*, **17**, 1365–1374.

Baker, G. A., Jr., Rushbrooke, G. S., and Gilbert, H. E. (1964), "High temperature series expansions for the spin-$\frac{1}{2}$ Heisenberg model by the method of irreducible representations of the symmetric group," *Phys. Rev.*, **135**, A1272–1277.

Domany, E., Mukamel, D., and Fisher, M. E. (1976), "Destruction of first order transitions by symmetry breaking fields," *Phys. Rev. B*, **15**, 5432–5441.

Domb, C. and Sykes, M. F. (1961), "Use of series expansions for the Ising model susceptibility and the excluded value problem," *J. Math. Phys.*, **2**, 63–67.

Essam, J. W. and Fisher, M. E. (1963), "Padé approximant studies of the lattice gas and Ising ferromagnet below the critical point," *J. Chem. Phys.*, **38**, 802–812.

Ferer, M., Moore, M. A., and Wortis, M. (1971), "Some critical properties of the nearest-neighbor, classical Heisenberg model for the fcc lattice in finite field for temperatures greater than T_c," *Phys. Rev. B*, **4**, 3954–3963.

Fisher, M. E. (1967), "The theory of equilibrium critical phenomena," *Rept. Prog. Phys.*, **30**, 615–731.

Fisher, M. E. and Burford, R. J. (1967), "Theory of critical point scattering and correlations I. The Ising model," *Phys. Rev.*, **156**, 583–622.

Gaunt, D. S. and Guttmann, A. J. (1974), "Asymptotic analysis of coefficients," in C. Domb and M. S. Green (eds.), *Phase Transitions and Critical Phenomena*, Vol. 3, Academic Press, New York, pp. 181–245.

Gaunt, D. S. and Guttmann, A. J. (1974), "Asymptotic analysis of coefficients," in C. Domb and M. S. Green (eds.), *Phase Transitions and Critical Phenomena*, Vol. 3, Academic Press, New York, pp. 181–245.

Griffiths, R. B., "Rigorous results and theorems," in C. Domb and M. S. Green (eds.), *Phase Transitions and Critical Phenomena*, Vol. 1, Academic Press, London, pp. 7–109.

Guttmann, A. J., Thompson, C. J., and Ninham, B. W. (1970), "Determination of critical behaviour from series expansions in lattice statistics, IV," *J. Phys. C*, **3**, 1641–1651.

Hunter, D. L. and Baker, G. A., Jr. (1973), "Methods of series analysis I," *Phys. Rev. B*, **7**, 3346–3376.

Moore, M. A., Jasnow, D., and Wortis, M. (1969), "Spin-spin correlation function of the three-dimensional Ising ferromagnet above the Curie temperature," *Phys. Rev. Lett.*, **22**, 940–943.

Nagle, J. F. and Bonner, J. C. (1970), "Numerical study of Ising chain with long range ferromagnetic interactions," *J. Phys. C*, **3**, 352–365.

Selected Bibliography on Padé Approximants and Electrical Network Theory

Brophy, F. and Salazar, A. C. (1973), "Considerations of the Padé approximant technique in the synthesis of recursive digital filters," *IEEE Trans. Audio Electroacoust.*, **Au-21**, 500–505.

Chui, C. K., and Chan, A. K. (1979), "A two sided rational approximation method for recursive digital filtering," *I. E. E. E. Trans. Acous. Speech and Signal Proc.*, ASSP-27 141–145.

Chui, C. K., Smith, P. W., and Su, L. Y. (1977), "A minimization problem related to Padé synthesis of recursive digital filters," in E. B. Saff and R. S. Varga (eds.), *Padé and Rational Approximation*, Academic Press, New York, pp. 247–256.

Fahmy, M. F. (1980), "The use of Padé approximants in the derivation of low-pass filters with simultaneous flat amplitude and delay characteristics," *Circuit Theory*, **8**, 197–204.

Hastings-James R. and Mehra, S. K. (1977), "Extensions of the Padé approximant technique for the design of recursive digital filters," *IEEE Trans. Assp.*, 501–509.

Scanlan, J. O. (1973), "Circuit theory," in P. R. Graves-Morris (ed.), *Padé Approximants*, Institute of Physics, Bristol, pp. 101–111.

Scott, R. E. (1977), "Low order rational approximations to the delay operator," *IEEE Trans. Ind. Electron. and Control Instrum.*, **IECI-24**, 61–64.

Shamash, Y. (1975), "Linear system reduction using Padé approximation to allow retention of dominant modes," *Int. J. of Control*, **21**, 257–272.

Sobhy, M. I. (1973), "Applications of Padé approximants in electrical network problems," in P. R. Graves-Morris (ed.), *Padé Approximants and Their Applications*, Academic Press, London, pp. 321–336.

Bibliography on Padé Approximants for Special Functions and Intrinsic Computer Functions

Fike, C. T. (1968), *Computer Evaluation of Mathematical Functions*, Prentice-Hall.

Gautschi, W. (1975), "Computational methods in special functions—a survey," in R. A. Askey (ed.), *Theory and Application of Special Functions*, Academic Press, New York, pp. 1–98.

Kogbetlianz, E. G. (1960), "Generation of elementary functions," in A. Ralston and H. S. Wilf (eds.), *Mathematical Methods for Digital Computers*, Wiley, pp. 7–35.

Luke, Y. L. (1964), *The Special Functions and Their Approximations*, Vol. I, Academic Press, New York.

Luke, Y. L. (1969), *The Special Functions and Their Approximations*, Vol. II, Academic Press, New York.

Maehly, H. J. (1960), "Rational approximations for transcendental functions," in *Proceedings of the International Conference on Information Processing*, Butterworth, London, pp. 57–62.

Bibliography on Hermite-Padé approximation

Given at least two functions $f_i(x)$ with power-series expansions

$$f_i(z) = \sum_{j=0}^{\infty} c_j^{(i)} z^j, \qquad i = 1, 2, \ldots, n,$$

with $f_i(0) \neq 0$ for $i = 1, 2, \ldots, n$, the Hermite–Padé problem consists of finding polynomials $Q_i(x)$ of maximal degrees σ_i, respectively, such that

$$\sum_{i=1}^{n} f_i(x) Q_i(x) = O(z^{\sigma})$$

where $\sigma = n - 1 + \sum_{i=1}^{n} \sigma_i$. See Section 1.3.

de Bruin, M. G. (1974), "Generalized continued fractions and a multi-dimensional Padé table," Ph. D. thesis, University of Amsterdam.

de Bruin, M. G. (1978), "Convergence of generalized C-fractions," *J. Approx. Theory*, **24**, 177–204.

Chudnovsky, G. V. (1980), "Padé approximation and the Riemann monodromy problem," in C. Bardos and D. Bessis (eds.), *Bifurcation Problems in Mathematical Physics and Related Topics*, D. Reidel, Boston, pp. 449–510.

Della Dora, J. and Di Crescenzo, C. (1979), "Approximation de Padé–Hermite," in L. Wuytack (ed.), *Padé Approximation and Its Applications*, Springer Lecture Notes in Mathematics, No. 765, Berlin, pp. 88–115.

Hermite, Ch., (1917), "Sur quelques approximations algébriques," *Oeuvres*, t. III, 146–149; "Sur la fonction exponentielle," *Oeuvres*, t. III, 150–181; "Sur la généralisation des fractions continues algébriques," *Oeuvres*, t. IV, 357–377, Gauthier-Villars, Paris.

Mahler, K. (1968), "Perfect systems," *Compositio Math.*, **19**, 95–166.

Padé, H. (1894), "Sur lagénéralisation des fractions continues algébriques," *J. de Math. Pures et Appl.*, **10**, 291–329.

Bibliography on Padé Approximants, Quasianalytic Functions, and Noise

Froissart, M. (1969), private communication to J. L. Basdevant, in "Padé approximants," in M. Nikolic (ed.), *Methods of Subnuclear Physics*, Gordon and Breach, pp. 1–60.

Gammel, J. L. (1973), "Effect of random errors (noise) in the terms of a power series on the convergence of the Padé approximants," in P. R. Graves-Morris (ed.), *Padé Approximants*, Inst. of Phys., Bristol, pp. 132–133.

Gammel, J. L. (1974), "Continuation of functions beyond natural boundaries," *Rocky Mtn. J. Math.*, **4**, 203–206.

Gammel, J. L. and Nuttall, J. (1973), "Convergence of Padé approximants to quasi-analytic functions beyond natural boundaries," *J. Math. Anal. Appl.*, **43**, 694–696.

Gilewicz, J. (1978), *Approximants de Padé*, Springer Lecture Notes on Mathematics, No. 667, pp. 309, 396.

Pindor, M. (1979), "Padé approximants and rational functions as tools for finding poles and zeros of analytical functions measured experimentally," in L. Wuytack (ed.), *Padé Approximation and Its Applications*, Springer Lecture Notes in Mathematics, No. 765, pp. 338–347.

Selected Bibliography on Assorted Applications of Padé Approximants and Related Topics

Fluid Flow

Van Dyke, M. (1958), "The supersonic blunt body problem—review and extension," *J. of Aerospace Sciences*, **25**, 485–496.

Van Tuyl, A. H. (1971), "Use of Padé fractions in the calculation of blunt body flows," *AIAA J.*, **9**, 1431–1433.

Van Tuyl, A. H. (1973), "Calculation of nozzle flows using Padé fractions," *AIAA J.*, **11**, 537–541.

Point Kinetics

Devooght, J. and Mund, E. (1975), "New developments in three-dimensional neutron kinetics and review of kinetic benchmark calculations," in *Proc. of Specialists Meeting, Garching Conference*, Nucl. Energy Agency Committee on Reactor Physics.

Indefinite Inner-Product Spaces

Van Rossum, H. (1953), *A Theory of Orthogonal Polynomials Based on the Padé Table*, Van Gorcum, Netherlands.

Van Rossum, H. (1979), "Orthogonal expansions in indefinite inner product spaces," in L. Wuytack (ed.), *Padé Approximation and Its Applications*, Springer Lecture Notes in Mathematics, No. 765, Berlin, pp. 172–183.

Calculation of the Matrix Exponential

Fair, W. and Luke, Y. L. (1970), "Padé approximants to the operator exponential," *Num. Math.*, **14**, 379–382.

Moler, C. and Van Loan, C. (1978), "Nineteen dubious ways to compute the exponential of a matrix," *SIAM Review*, **20**, 801–836.

Wragg, A. and Davies, C. (1975), "Computation of the exponential of a matrix II. Practical considerations," *J. Inst. Math. Appl.*, **15**, 273–278.

Molecular Polarizability

Barnsley, M. F. (1975), "Bounds on dynamical polarizabilities at high frequencies," *Molecular Physics*, **29**, 1377–1386.

Corcoran, C. T. and Langhoff, P. W. (1977), "Moment theory approximations for non-negative spectral densities," *J. Math. Phys.*, **18**, 651–657.

Langhoff, P. W. and Karplus, M. (1970), "Application of Padé approximants to dispersion force and optical polarizability computations," in G. A. Baker, Jr., and J. L. Gammel (eds.), *The Padé Approximant in Theoretical Physics*, Academic Press, New York, pp. 41–97.

Percolation

Dunn, A. G., Essam, J. W., and Ritchie, D. S. (1975), "Series expansion study of the pair connectedness in bond percolation models," *J. Phys. C*, **8**, 4219–4235.

Grommer Functions

Edrei, A. (1975), "The complete Padé tables of certain series of simple fractions," *Rocky Mtn. J. Math.*, **5**, 559–584.

Solitons

Lambert, F. (1980), "Padé approximants and closed form solutions of the KdV and MKdV equations," *Zeit für Physik (Particles and Fields)*, **5**, 147–150.

Turchetti, G. (1980), "Padé approximants and soliton solutions of the KdV equation," *Lett Nuovo Cim.*, **27**, 107–110.

Turbulence

Kraichnan, R. H. (1970), "Turbulent diffusion: evaluation of primitive and renormalised perturbation series by Padé approximants and by expansion of Stieltjes transforms into contributions from continuous orthogonal functions" in G. A. Baker Jr. and J. L. Gammel (eds.), The Padé approximant in theoretical physics, Academic Press, New York, pp 129–170.

Index for Part I and Part II